ROBERT WHITMAN
LOCAL REPORT

Julie Martin, Editor

Experiments in Art and Technology | Lafayette College Williams Center Art Gallery

Robert Whitman Local Report has been made possible by generous support from the Baker Family, with additional support from Experiments in Art and Technology, and the Williams Center Art Gallery, Lafayette College. It has been published in conjunction with the exhibition *Robert Whitman Local Report*, organized by the Williams Center Art Gallery, Morris R. Williams Center for the Arts, Lafayette College, Easton, Pennsylvania, held 18 February–25 March 2007.

Local Report, 2005, was sponsored by the Baker family, with additional support from National Realty & Development Corporation; Nokia, Connect to Art Initiative; Experiments in Art and Technology; Skidmore, Owings & Merrill Education Lab; and Dia Art Foundation.

The Williams Center Gallery program is funded in part by the Pennsylvania Council on the Arts—a state agency funded by the Commonwealth of Pennsylvania—and the National Endowment for the Arts.

Published by Experiments in Art and Technology, Berkeley Heights, New Jersey 07922, and the Lafayette College Williams Center Art Gallery, Easton, Pennsylvania 18042.

ISBN: 987-0931286-09-4
Library of Congress Control Number: 2007935156

Contents

Acknowledgments

On behalf of the Williams Center Gallery, Robert S. Mattison, Metzgar Professor of Art, and Paul Miller, visual resources curator, department of art, I wish to thank Robert Whitman for the opportunity to present *Local Report* at Lafayette College. We were privileged to be active participants in the creation of the piece; thus, we are especially pleased to show *Local Report* in Easton, one of the communities in which a performance took place. In addition to having the opportunity of seeing the Easton *Local Report* in the context of the complete project, local participants can now enjoy the results of their efforts, something they were unable to do while "reporting."

In late spring 2005, to determine whether the Easton area was an appropriate venue for his newest telecommunications project, Robert Whitman, and Julie Martin, the project coordinator, contacted Bob Mattison. The response was positive; Bob enlisted the help of Paul Miller and me. As the summer progressed, we became increasingly excited about our involvement in such an ambitious and intriguing project—a project that, as it unfolded over five successive weekends, would create video and audio portraits of Easton; Trumbull, Connecticut; Holmdel and Burlington, New Jersey; and Kingston, New York.

Mattison and I served as the local coordinators, charged with finding the 30 reporters, and Miller became an important member of the production staff by providing technical assistance for the Easton setup at Northampton Crossings. Because Easton's *Local Report* performance took place before the start of the fall semester, we were unable to enlist Lafayette students, but we did succeed in recruiting Lafayette faculty, staff, and alumni. The local arts community responded enthusiastically, with many coming from ACE, the Arts Community of Easton. Coincidentally, one performer, Bruce Wall, had participated in Whitman's projection piece *Palisades* in the late seventies.

As longtime Easton residents, Mattison, Miller, and I showed Julie Martin and Bob Whitman locations that are uniquely Easton—Center Square, Mount Ida, Easton Cemetery, the Delaware and Lehigh rivers, Bushkill Park, and, of course, the Lafayette College campus. Reporters were sent to these locations as well as other areas in the Lehigh Valley and western New Jersey. Some of the video that the reporters chose to submit came as a surprise; even though we had selected many of the locations in advance, we could not have anticipated the visual and auditory decisions the reporters ultimately made. As a result, we discovered new things about our community and fresh ways of viewing the familiar. The videos, taken with cell phones and enlarged many times in the exhibition projections, became blurry and jerky, lacking the pixels for high definition and clarity, and yet they possessed a dreamlike, timeless quality. Sometimes, it is difficult, if not impossible, to identify what was filmed, but careful viewing reveals common themes as well as unique perspectives.

I thank the Baker family for supporting the publication; Julie Martin, project manager, for her invaluable diligence and considerable efforts in overseeing the many details of the exhibition and catalogue; Hedi Sorger for her work on the catalogue; and Bettina Funcke for her informative and insightful essay. The installation would not have been possible without the loan of computers and the technical assistance of Rich Santillo, of Lafayette's department of information technology.

My particular thanks to Robert Whitman. He was a pleasure to work with. His innovative concepts and artistic genius, evident in such works as *Local Report,* encourage us to consider the creative potential of seemingly simple communications tools in exciting new ways.

Michiko Okaya
Williams Center Art Gallery Director
Lafayette College
Easton, Pennsylvania

4

Preface

When you go to the web site for the National Realty and Development Corporation, you first encounter a series of quotes from such artists as Edgar Degas and Andy Warhol on the power and the value of art and artists to society. These quotes reflect the Baker family's longtime commitment to the arts. Richard Baker has realized, for a number of years, the importance of introducing the work of leading contemporary artists to communities that he and his company are involved in. By sponsoring artists to work on special projects at NRDC-owned shopping centers, he has introduced not only the shoppers but also the whole community to art projects that both stimulate and enlighten. The public's positive response is clear evidence that people of all ages and circumstances value Baker's efforts, and that exposure to the artists often enriches their lives. As Baker himself has said: "If we are able to see different types of work, and appreciate it no matter where we live, it opens our minds and influences our lives and makes our communities richer, more exciting, and special places to be."

This spirit and vision have led Richard Baker and the Baker family to sponsor the five performances that make up *Local Report*, to generously donate the final installation of *Local Report* to the Solomon R. Guggenheim Museum, in New York, and, now, to make this catalogue possible. For their inspiration, support, and encouragement, we are most grateful.

A tremendous undertaking, *Local Report* relied on the expertise of many individuals. The production team was extraordinary: In organizing and carrying out the logistics of this operation, they were able to produce a performance a week, for five weeks, in widely separated venues. I am pleased that some members of the team added their voices to tell the story of the project for this catalogue.

Local Report was a unique collaboration with 150 people, in five different communities, who acted as the Local Reporters. By engaging in the artistic process, they were encouraged to use the new technology to create portraits in unique ways.

Hedi Sorger, Michiko Okaya, and I worked with Whitman to develop a layout and design that would translate the flow of voices and video images of the piece into the linear text and still images of the printed page. For each venue, we transcribed all of the voice calls and reproduced one image from each reporter's video.

With this catalogue, we wanted not only to record and document the performances of *Local Report*, in the summer of 2005, but to point the way to the future: In the last section of the book, Shawn Van Every and Hans-Cristoph Steiner have written a How-To for anyone interested in producing their own Local Report. Bob Whitman's work is not only art history but may well prove a significant part of art's future.

Julie Martin
Local Report Project Manager

Local Report *production crew Northampton Crossings, Easton, Pennsylvania, from left: Robert Whitman, Paul Miller, Hedi Sorger, Shawn Van Every, Anne-Olivia Le Cornec, Kieran Sobel, Hans-Christoph Steiner, Julie Martin, and Walter Patrick Smith.*

Hedi Sorger and Robert Whitman putting up a poster at Northampton Crossings, Easton, Pennsylvania.

The tent housing Local Report *performance venue at the shopping center, Northampton Crossings, Easton, Pennsylvania.*

The Story of *Local Report*   Julie Martin

Local Report is a multiscreen video and sound installation by Robert Whitman made up of five live performances, from 2005, in which participants were asked to use video cell phones to transmit images and sounds from their communities.

The performances for *Local Report* were sponsored by the Baker Family, in conjunction with the National Realty & Development Corporation, with the purpose of providing an opportunity for shoppers at the NRDC shopping centers to experience works of art. The performances were held at five NRDC shopping centers in New York, New Jersey, Pennsylvania, and Connecticut, on five weekends between July 29 and August 28, 2005, in vacant stores or in tents in the parking area of the shopping center. The performances were streamed live to the project's web site: www.whitmanlocalreport.net.

Richard Baker and the Baker family donated a copy of *Local Report* to the Solomon R. Guggenheim Museum, in New York. The work was installed there on December 5, 2005, at an event marking its formal accession into the permanent collection of the museum.

In January 2006, *Local Report* was installed at an exhibition at PaceWildenstein, using video and sound from three of the five venues. The exhibition at the Williams Center Art Gallery at Lafayette College, in March 2007, presented all five venues in an installation designed by Whitman in collaboration with gallery director Michiko Okaya.

This is the story of these performances, which created *Local Report*, from interviews with Robert Whitman and some members of the project team: Anne-Olivia Le Cornec, Hans-Christoph Steiner, Shawn Van Every, Walter Patrick Smith, and the author.

WALTER PATRICK SMITH: When I met Bob Whitman in early 2004, we talked about the SOM [Skidmore, Owings & Merrill] Education Lab and its ongoing commitment to working with artists on our school projects, and our ideas about how to make schools better by changing how schools teach—not just the teachers but how the school buildings themselves can teach. Bob told me about some of his earlier projects with students using telecommunication equipment to communicate between schools, and how important it was for them to be able to reach out. When we won the commission to work on the Creative and Performing Arts High School, in Camden, New Jersey, we thought that Bob would be a natural to help us on this project.

During our drives to Camden, Bob and I talked about lots of things, from Sophocles to Fellini's clown film. Then, Bob introduced me to this piece of his called *Architecture*, in which he had people wandering through an old pier in Manhattan. It was about the importance of the place, and leading each other around, and showing each other places.

I had been working for several years with Richard Baker on school projects and art projects for his home and for his shopping centers. So, as I got to know Bob, we brought him out to take a look at Richard's woods; and Richard commissioned Bob to create a work of art for his forest, combining sculpture and projected images.

Then, I went to see Bob's theater performance *Antenna*, in Leeds, England, in the fall of 2004. I said to Richard, "Could we do a Bob performance? Let me tell you these things we've been talking about: We have the same fascination with clowns and his sense of timelessness; maybe we could do kind of a gypsy thing with your malls." Richard loved it; it was about his malls. He immediately said, "Let's do it."

JULIE MARTIN: Bob proposed making a series of performances, which he subsequently titled *Local Report*. It was a new version of a work he had performed several times in the past, both updated and specially designed for the shopping center environment.

ROBERT WHITMAN: I did the first one in 1972. Somebody asked me to participate in a series of radio programs on artists. I thought, O.K., radio is such an un-

7

derutilized medium, it'll be interesting to use the radio in a way that I thought would be surprising to me. What I did was set people up with coins, sent them around to particular pay phones in Manhattan, and have them move from phone to phone and make a report at each one about what they saw when they called and got on the air. This was broadcast live on WBAI. We ended up with a collection of these verbal images of Manhattan and lots of funny, surprising coincidences. You know, fire trucks coming across the sound in one case, and then the same fire trucks as they moved five blocks to where the next guy happened to be, or something like that.

That was the first one: I called it *News*. After that, I did the same piece in several different cities—Houston, Minneapolis-St. Paul, and so forth. In each of those cities, I got to have a tour of the environment from someone who was a fan of the environment. I got to see things you would never get to see otherwise, and in a context you would never experience, because people aren't usually going to take you on a tour of, let's say, the railroad switching yards at night. So, you get all these surprising, and sometimes wonderful, experiences.

I've always thought that it would be a nice thing to do—and I've tried to do it a couple of times—is have a piece change over time and develop itself and go through different configurations, and still keep what I call the essential essence of a piece. So, this piece developed along those lines.

As the technology changed, you have the cell phone: You don't need to have a bunch of dimes or nickels or quarters anymore, you can call from anywhere in the city! We did the piece in 2002, in Leeds, using cell phones, and the reports played over loudspeakers in a large square in the middle of town.

Finally came the video cell phone. This latest incarnation uses video cell phones, and the reporters make both video reports and voice calls. As it turned out, the broadcast in this case was on the Internet.

What these pieces have in common is that you get to see something that is generated by a community, you get to see what other people see, and have a kind of surprise in all that. It's just that we're able to change the piece by using this new technology.

WPS: Bob had this vision of what he wanted to do. It was interesting, the notion of suddenly taking the common man and making that person the main performer. But I was really shocked the first time we talked about it. I was, like, this isn't what I was thinking about at all. What did I get myself into here? I don't know anything about working on one of these performances. And I had thought there were going to be some jugglers who showed up in the parking lot doing tricks.

Quickly, Bob came up with how, if we handed out 30 video cell phones and each person did a 20-second clip, that would turn into ten minutes, and then, if we had them each do three clips, it would be a half hour. We'd do this show for a half hour.

I thought, at first, that the performance could travel to all of Richard's shopping centers; I have a really bad sense of reality at times: He has more than 70 shopping centers. But Bob quickly trimmed it down to five locations. Then, we talked about how the five should be as dispersed as possible, that we would pick locations in five states. We wanted to have as wide a cast as possible to be able to record the cultural differences between the way people would talk and what they would look at in different communities.

We began to organize the five events. There would be 30 reporters who would make video and voice reports. The video reports were going to be projected on a screen and the voice reports would be played over speakers.

JM: The first idea was to locate the performance in the middle of the parking lot, and all the equipment could be in a tractor-trailer with the projection screen on the side or inside. We also considered renting a RV, so all the equipment and control panels could be inside, and we could just drive up, park in the parking lot, erect a screen, and run the performance from the RV. We soon decided, however, that the performance sites should be in empty stores at the shopping centers, where the equipment could be installed, the video could be projected, and people could gather.

WPS: During one of our talks with Richard, he said we should do this during the summer, starting in

August, to get it done before school starts—and this was in May. It was an ambitious schedule.

JM: I had been meeting with Bob and Walter, and realized that E.A.T could be useful in organizing some aspects of the technical systems that Bob wanted to use. So, E.A.T. staff members Hedi Sorger and Anne-Olivia Le Cornec joined me on the team.

The first task was to choose the five shopping centers to be the sites of the performances. NRDC has an excellent web site of its properties, showing location, size, and demographics of the surrounding area and the availability of empty stores to be the performance headquarters. We looked for shopping centers in five different states, and for centers that had fairly large populations surrounding them. Since we had a limited time to find the 30 participants, we chose areas where we had some contacts with local people or institutions. We chose Northampton Crossings, in Easton, Pennsylvania, because we knew Robert S. Mattison in the art department at Lafayette College; and we stuck with it, even though, as it turned out, there was no empty store available. Ultimately, we held the Easton performance in a large tent erected on the parking lot. Hawley Lane Plaza was in Trumbull, Connecticut, close to Yale

Outside speakers and banners identify the Local Report *venue at Liberty Square Center, Burlington, New Jersey.*

Hedi Sorger puts up posters at the Local Report *venue at Hawley Lane Plaza, Trumbull, Connecticut.*

University and Bridgeport, where Whitman had some artist friends; Holmdel, New Jersey, was close to Bell Laboratories and had some active junior colleges and arts organizations; Kingston, New York, was chosen for its proximity to Bard College and Pauline Oliveros's Deep Listening Institute; Burlington, New Jersey, was chosen when two other state locations—Northampton, Massachusetts, and Brattleboro, Vermont—were impossible for technical reasons. Burlington turned out to be close to both Princeton University and Philadelphia, and we found participants from both places.

The choice of shopping centers also depended on whether we could have cell-phone coverage in the area. As the performance system developed, we also needed to have high-speed DSL or cable connections to the Internet.

The shopping centers we chose were not in any major urban areas, and it soon became obvious that both cell phone service and Internet access was not a simple matter. These networks were not as ubiquitous or as uniformly strong and reliable as we city dwellers had experienced. Hedi Sorger spent several hours on the

Internet, looking at the cell phone providers' less-than-detailed maps of their coverage areas. After long conversations with technical representatives of the different companies, she detemined that only Cingular and T-Mobile would cover the areas we needed, and that Cingular was the best overall.

WPS: We started to meet at SOM's offices, and began to look at the cell phones everyone there had. We all had different ones; they all did different things. My colleague Peter Cho had a video cell phone, and, after working with it for a while, we realized that the phones would have to be adapted to what Whitman wanted to do with them.

JM: I began to look for a technical person who could work with us on adapting the cell phones. Through Sue Wrbican, a video artist who had worked for E.A.T., we found Shawn Van Every, researcher and adjunct professor, at the Interactive Telecommunications Program of New York University, and Shawn brought in Hans-Christoph Steiner, who had studied at ITP and was working with computer music, open-source software programming, and performances with collaborative artists groups in the city.

SHAWN VAN EVERY: I met with Julie and Walter first, and then Bob came the next time to NYU. They described something Bob had done before, where people could call the radio station live from pay phones around the city and describe what they saw. Now, you guys wanted to use video off the phones to do the same thing. That was something that was very appealing to me, combining citizen journalism with mobile phone video, allowing anyone to become a reporter, on par with traditional media. It was something that I was actually working toward in my own work and with my students. So, I demonstrated for Bob and Walter an application that I had written for a cell phone, which shot live video and then automatically uploaded it to a video blog, a web site on the Internet. That seemed to be exactly what you guys needed. It needed some hefty work, but the concept was there.

Shawn Van Every, Walter Patrick Smith, Hans-Christoph Steiner at a planning meeting at New York University's Interactive Telecommunications Program.

JM: Whitman wanted both video and audio, and requested that the sound and video should not be related to each other. The systems needed to be designed so the two components came in on separate tracks.

SVE: Bob had an idea of what he wanted, and he wanted to know what could be done. It wasn't terribly difficult to explain it at a high level to him. Once he gave the go ahead for doing things in the way we thought it could be done, it was pretty straightforward.

HANS-CHRISTOPH STEINER: It seemed to me, in the beginning, he asked questions about what was possible; he cared a lot about what was possible. Once we had laid out what we were doing, it didn't matter to him how it was implemented, as long as it worked. After that, it was, "O.K., this is your bit; go to it."

JM: We had several meetings at NYU. Van Every had outlined his plan for getting video from the mobile phones to the Internet and then to the performance sites. Van Every's proposal to work with the Internet made it possible to use cell phones to gather the voice and video reports and play them at the performance site, but also to stream them in real time back over the Internet.

RW: I always thought it was important to have broadcast possibility of the piece. The Internet was the only easily available broadcast situation, and anybody who wanted to watch it could, during the performance and at any time afterwards.

JM: Since Nokia phones would work with the Cingular network and Van Every was familiar with the Nokia environment, we approached Nokia for support of the project. Jean Pierre Lespinasse, head of the Connect to Art Initiative at Nokia, quickly agreed to support the project and provided us with 33 Nokia 7610 video cell phones. Van Every and Steiner were able to begin writing applications.

SVE: Every mobile phone has its own quirks, every brand and every model, unfortunately. We were working with limited memory, limited network conditions. The heavy work was creating an application that would be fairly fault-tolerant, that would work in a wide variety of conditions on the Nokia phones that we had chosen. Also, I had to make the phone easy to use, so you didn't have to be a programmer in order to use it.

Shawn Van Every sketching a diagram of the overall system architecture for Local Report.

What I ended up doing was having a one-button interface that would shoot 20 seconds of video, and then automatically upload that video to an Internet web site. At each location, we had a laptop hooked up to a projector that ran an application to play back the video clips full screen as they came in from the server.

Bob also wanted the ability to make voice calls in real time, so that's where we went next. Hans and I had been playing a little bit with something called Asterisk, which is an open source PBX, for multiple telephone lines. Hans had actually gone a lot further with it than I had, so I turned to him to help me and do a lot of the phone work.

HCS: Asterisk is free software, a software version of an office phone system. Regular office phone systems you get through the phone company are extremely expensive, but with Asterisk you can run a really large phone system with many lines on a PC that someone just threw out on the street.

SVE: Essentially what Hans created was an on-hold system like some help line would have. There was a queue, there were audio prompts, and there was music they could listen to while they were in the queue. All the people who were participating in the event would be able to call up, wait in line. Then Bob, who was sitting at a master control terminal, which was running a piece of software that could receive the phone calls from the Internet phone server, would hit the button, and their voice would then come into the performance.

HCS: We decided to have five phone lines, based on the fact we had 30 people. It was basically to make sure we had enough of a queue, so when people called, they wouldn't get a busy signal but would be put on hold. Each line could support four simultaneous calls, so five times four is 20—and, theoretically, more than 20 people wouldn't be calling at any one time. I think, at the peak, we had a maximum of 14 people on hold.

We needed a way to let people know they were still on hold. We decided to play a piece of music that was solid so you could hear the rhythms. I tried something of mine, but it really didn't work on the phone that well, so we ended up playing Terry Riley's *Olson III*. We also had these short prompts of recorded voices. The final one said, "Get ready." Then, when the music stopped, the person was supposed to start speaking.

SVE: Our system for *Local Report* was like a radio station, because we had people calling up the phone system. They were queued up, and when we hit the button, they were being broadcast to the speakers and streamed at the same time.

HCS: The key part of the system was that we were using no actual phone lines; it was all Internet-based. So, we have the server, and then we got phone service from Firefly, a VoIP (voice over Internet protocol) provider. They did all the interfacing on the Internet, and all we had to do was plug into the Internet. We could get a total phone system for wherever we happened to be, a mall parking lot or an empty store. We could use this system anywhere there was an Internet connection.

JM: Hedi began to call to Verizon and other local phone companies in the five areas to determine whether

Anne-Olivia Le Cornec greets Local Reporters arriving at Liberty Square Center, in Burlington, New Jersey.

Anne-Olivia Le Cornec and Robert Whitman studying maps to assign reporting locations.

we could get the high-speed and reliable DSL Internet connection that Shawn and Hans asked for. We needed approximately three megabits downstream and approximately one megabit upstream, for the streaming video. We could have gotten away with less, but the videos would have come in slower and the stream of the event wouldn't have been good quality. The performance site had to be less than 10,000 feet from a telephone company's CO, or central office, to get the speeds we required. We were able to have a phone line and DSL modems installed, and have sufficiently robust DSL service to the centers in Trumbull, Holmdel, Burlington, and Kingston. At Northampton Crossings, in Easton, we declined the cable company's offer to tear up the asphalt to install cable connections to our location, and then we were able to order a T1 line connected to the telephone box in a utility room at the back of the nearest store—a bridal boutique. The NRDC staff at Northampton Crossings did an elegant job of stringing the cable for power and Internet from the store, over a roadway, and down into our tent.

JM: We soon found out that we were not the usual customer when inquiring about cell phone and Internet service: Our requirements for cell phone service and DSL Internet connections did not match what the companies were offering. They, of course, want long-term contracts, and here we were wanting to use 30 phones for one half-hour each, once a week, for five weeks—between July 29 and August 26. And we needed a phone line and modem for DSL Internet connections installed either in empty stores or in a tent in parking lots for the period of one to two months at the most. We calculated the number of minutes the whole project required, and then bought Cingular Wireless Pooled Business Service with five meg Media Net for sending video for 30 phones for two months. We got our DSL service through the local Verizon phone companies.

JM: For each of the locations, we found a person to act as Local Coordinator, who would recruit the 30 Local Reporters for the performances and help in the set-up. In particular, Lisa Barnard, in Kingston, oversaw the installation of the DSL line in the abandoned office of a car repair garage at the shopping center. The Local Coordinator also organized a tour for Bob of the area surrounding the shopping center, to scout locations for the reports.

RW: The shopping centers that we picked were in places that I hadn't really been, so I got a chance to get

Greeting the Local Reporters at Hawley Lane Plaza, Trumbull, Connecticut: Hedi Sorger, Julie Martin, Cynthia Rubin, and Robert Whitman. Shawn Van Every adjusts the video projector.

taken around and be shown the sights in different places. In Burlington, I was taken around by a local builder who specializes in restoring the colonial-era houses in the town; in Kingston, a local newspaper editor was my guide. Michiko Okaya and Bob Mattison showed me the places they really liked in Easton, including a mausoleum with Tiffany glass windows in the local cemetery. My tour guide in Trumbull was a man named Mickey. The day we took the tour was his ninety-eighth birthday. The wonderful part of that tour was that it was almost like a ghost tour, because he remembered where everything had been. Oh, here was where the plane crash was in 1935, and I came out to see. And here was where the icehouse was . . . So, it's kind of spectacular, and, of course, none of those things are there; they all were there, but they're not now. And that was wonderful.

ANNE-OLIVIA LE CORNEC: I enlarged the maps of the areas, so Bob and I could easily find the specific places he wanted to send the Local Reporters. Bob had a predilection for train stations, rivers, bridges, big farms or orchards, towers or big antennas revealing technical installations, factories, memorials, and cemeteries. It was fun to explore the map and locate the places he remembered from his tours and that he wanted to assign to the Local Reporters. When the places were chosen, I made smaller detailed maps and instructions for about 90 places for each venue, three different places for each reporter.

RW: I wanted to keep the different reporters separated and spaced out a bit. Every area does have a particular landscape that's accessible to a lot of people. Places have little rivers, they all have parks, they all have graveyards. You've got to go for those. They have industry, you go for those. But it's basically just to keep people spread out, I think. They'd find different things to take pictures of even if they were all in the same place.

JM: The performances took place on Friday afternoons, from July 29 to August 26, and the resulting video and sound reports were played at the shopping center on the following Saturday and Sunday. During the five weeks of the performances, Anne-Olivia and I had a condensed version of the idea of the truck pulling into the shopping center parking lot. We rented an SUV, into which Anne-Olivia packed the video projector, the screen, the speakers, the sound mixer, wires for the audio, the wireless router, and cables for the Internet connection, etc. We could store the equipment in the car during the week, then drive to the designated shopping center on Friday morning, and unload in time for the setup.

Bob, and Shawn or Hans, and Walter, and the other crew members arrived, hooked up to the Internet connection, the wireless routers, and the laptop computers, set up the screen and the projector, set up the speakers, and got ready for the performance.

The sound system, in which speakers were inside the store and also outside for passersby to hear, was designed by Patrick Heilman of the Dia Art Foundation.

Robert Whitman talking to Local Reporters at Northampton Crossings, Easton, Pennsylvania.

Dia lent us the sound mixer and speakers. Patrick participated in the first venue, at Trumbull, Connecticut. For the following four performances, the sound equipment was organized and installed by Kieran Sobel, a college student who had wandered into the Trumbull performance site on Saturday while we were playing back the video and audio for the weekend. He helped us with a computer glitch, and was both interested and knowledgeable about computers and sound equipment. He was on summer break, and we recruited him on the spot. Paul Miller, at Lafayette College, worked on the sound and video installation at the Easton venue, particularly setting up the larger 10K lumens projector that we had for that site.

JM: The reporters were asked to arrive early in the afternoon. They were given maps and directions for three locations to go to, and picked up their cell phones. When all the reporters had arrived, Bob gave a brief introduction to the project, and some instructions about what he expected: In Easton, for example, he told the reporters, "My directions are quite simple: What I want is a video of what you see where you are. And for the verbal part: First, say where you are and then describe something, whatever strikes you in some way, very straightforward, the way it is." He also made it clear that it wasn't necessary to film and verbally describe the same thing. After Bob spoke, Shawn or Hans gave instructions in how to use the video and voice call functions on the cell phones.

SVE: That was fun. We had tried to make the interfaces as easy as possible, since mobile phones are sometimes difficult. The kids got it right away—but for older people, it was harder. We had some instruction sheets to hand out, and we had a big poster that had labels for all the different buttons and what they did. Before the performance, we explained that this is what this button does, this is what that button does. We told them that we were sorry we couldn't make this completely automatic, but that they were smart and could do it.

For the voice calls, we put the five phone numbers that went to the Internet site into the phone's address

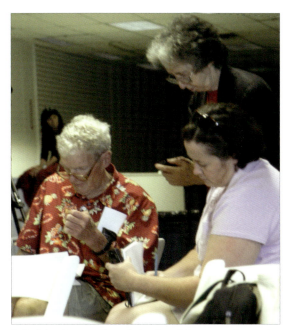

Julie Martin, standing, helps Local Reporters learn the cell phones at Kohl's Plaza, Holmdel, New Jersey.

book, and the reporters could dial any one of them by simply pressing one button.

For the video function, they pressed the button we had labled WLR, which brought up the video screen. When they pressed a button to start shooting video, the phone automatically recorded for 20 seconds. Then, they pressed one more button to send the video to the Internet web site. It took, maybe, a minute in the best case or, maybe, five minutes in the worst case, to upload that 20 seconds of video.

One of the problems we had during the first performance was that people would shoot video and begin to send it, but then they would get in their car and drive to their next location. They would lose the cell tower connection, and their video wouldn't finish being uploaded. The next performance, we instructed people to stay put after they shot the video until it was done uploading. That way, it worked much better.

HCS: There was another problem: the variation in the strength of the wireless connection and the time it took for the video to be sent. At the first performance, some people said, "I was standing there for ten minutes, and it was still sending." So, they quit the transmission.

SVE: We wrote some software so that, after five minutes, the uploading would quit by itself; then, they could shoot again. After the first venue, we started getting way more video than we could play in the half-hour period.

HCS: By the end, it was like a flood.

JM: For his earlier telephone pieces, Whitman had developed a script or score for the reporters, in which he "set up an ideal situation, or program, and then it got played like an instrument, almost. There were 30 reporters, which may be a couple or a single individual, and they go out and they make three reports from three different places within half an hour." He broke the reporters into two groups and the performance into three 10-minute time periods. He asked the reporters to start at five o'clock and make one voice and one video report, and move to their next reporting place within a 10-minute time period, and for the two different groups to stagger when, in the 10-minute period, they would make their reports.

AOLC: We found that broadband coverage, and, therefore, signal strength, was not uniform in different places around the area, and it could take between

Local Reporters learning to use the cell phone at Northampton Crossings, Easton, Pennsylvania.

one and three minutes or more for the video clips to be uploaded to the server. So, on the first weekend, we realized that the timing-direction rules that Bob wanted were not possible because of the signal strength and these coverage gaps. By the second weekend, we knew how to deal with this problem.

JM: While the video clips took several minutes to be uploaded to the Internet server, they could be stored on the server in the order they had been received. From the second weekend on, Whitman asked the reporters

Robert Whitman at the laptop controlling the voice calls during the performance at Northampton Crossings, Easton, Pennsylvania.

to go to their first locations at four o'clock and to begin sending video clips at four-thirty, so there would be a "queue" of video clips ready to play.

At five o'clock, Bob cued his fellow performer—sometimes Shawn, sometimes Hans—to start playing the video clips, which were sent to the video projector. The video clips then played automatically in the order they were stored on the server until the performance was complete, one 20-second video clip after another.

The reporters were instructed to begin their first voice call at five o'clock.

RW: When the piece was first done on the radio, someone there told the person calling up that they were on the air. For *Local Report,* we had hold music by Terry

Riley, which was kind of cool hold music. They were instructed to start making their report when the music stopped. That way, no one's voice intruded to tell them they were going to be on the air.

JM: Whitman pushed a button on a VoIP software phone to put the reporter on live and route their voice over the speakers. Whitman "hung up" when he felt the caller had "created a coherent image." At this point, the next caller in the queue began to make his or her report. The voice reports continued in this manner throughout the performance.

RW: There's no editing: it is what it is. Nature's just fine. I think it's more valuable to have the surprise and not make something that has to do with your expectations of how it's supposed to be.

JM: The video clips and the voice reports were recorded and played back in separate but continuous loops immediately after the performance.

AOLC: After each performance, the reporters came back to the site. We had a buffet party while we looked at the loops of sound and video. Everyone was surprised to see how Bob made their everyday images into an artwork, and were happy to be part of the poetry.

WPS: The people came back and watched it, and they had great moments of joy when they saw theirs, or they wondered where someone else's image was taken. There was this ability for information to be shared and, in a common way, to create a sense of community.

RW: It's always interesting to me what someone is going to take a picture of. I have the view that there's no such thing as a bad picture . . . it's really hard for me to imagine one. The results are always surprising. They always took things that were unexpected. Well, not necessarily. If you're going to send somebody out by the water tower, they're probably going to take a picture of the water tower. But sometimes what they take is completely unexpected.

You can see, a lot of times, why they took a picture of something—they thought it was stunning. Other times, people took pictures of things that were just *there*. In a way, that's more fun; it's spontaneous, spotting something and taking a picture of it.

I should have asked what their experience was, but if they had told me, I would have modified the directions, according to the interaction. It's much better if people are free to take a picture of "that"—whatever it might be. If it's a duck, or a little flag fluttering somewhere, who knows . . . .

I also can't predict what people are going to say about what they are seeing. But if you listen to their voice reports, they're what I would call sensitive poetry. Actually, it's the sensitivity of the people making the reports that makes the poetry. The people go out there,

just get into the program, into the zone, and work with it; the results are wonderful. Just the simple act of describing something can be wonderful. Somebody says, "I'm looking across the street, at the traffic and there's a lady on the other side of the street, holding a brown paper bag." All of a sudden, you have this terrific picture of something . . . it's in your mind. With *Local Report*, you have a collection of these images. We end up getting a cultural image of the community in which the piece was made.

AOLC: On the Saturday and Sunday that followed each performance, the video and audio loops of the reports were shown at the performance site in the shopping center. We had all kinds of people stopping by to look at the video, recognizing their area, and wondering what was going on. Many expressed regret at not

From left: Robert Whitman, Kieran Sobel, Hans-Christoph Steiner, and Walter Patrick Smith at the storefront control center during the performance at Kingston Center, Kingston, New York.

having been part of it. They liked the concept of the performances. Some called their friends or family, telling them to go and see us in the following towns.

JM: Bob had decided he would incorporate each of the performances into a final five-screen video-and-sound installation. For this multiscreen installation of *Local Report,* the five videos from each site play on five screens, and the sound from each venue is localized to the screen showing the video from its venue. The video images and the sound play on different tracks; because of their varying lengths, the sound shifts in relation to the image as the work is shown.

RW: I like the idea of making a work that is unregulated and completely undetermined by the way it's set up. Because it has a form that's unique to the piece, it suddenly exists on its own terms, no matter how you do it. My own choice is to not exercise control. I like the poetry in democracy. You get a huge variety of inputs from all these unexpected sources, which nobody can be in control of. It's also nice when someone does something on their own, when they don't necessarily heed the specific directions. You always have to accept the sheer wonderfulness of somebody's understanding. It's not all like an even peanut-butter spread of material.

Richard Baker and Robert Whitman at Kohl's Plaza, Holmdel, New Jersey, performance.

AOLC: It was a good social experience. I loved the way Bob managed to bring people together to collaborate on an artwork that made them recognize the poetry in their everyday surroundings.

RW: When we got the phones, they were state of the art. The thing about these phones is that the format for video is unusual: The aspect ratio is different from normal video; the frame rate varies within the clips; the resolution is very low. Video phone technology has already changed and gotten better, so you'll never, ever get the same low-res character of these particular video reports again.

WPS: In 30 years, no matter how good the technology becomes, we'll think—what a remarkable time capsule this is, what people could think of, and what they were able to show pictures of. *Local Report* is an amazing story of these intangible things that we all share as groups and communities in the places where we live. It's probably the most timeless piece I can imagine creating.

Visitors to the shopping center watch the Local Report *performance at Hawley Lane Plaza, Trumbull, Connecticut.*

Installation of Local Report *at the Solomon R. Guggenheim Museum, New York, December 6, 2005.*

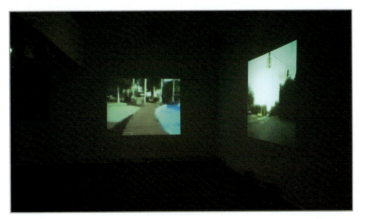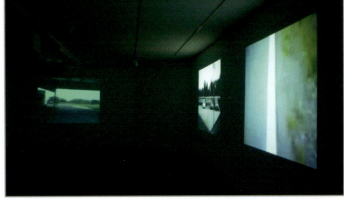

Local Report *installation at Lafayette College, Easton, Pennsylvania, 18 February–25 March, 2007; from left to right, Easton, Pennsylvania; Trumbull, Connecticut; Burlington, New Jersey; Holmdel, New Jersey; and Kingston, New York.*

Robert Whitman's Telecommunication Projects   Bettina Funcke

The work of Robert Whitman, dating mostly from the sixties and seventies, deserves serious reconsideration: It may be seen, in part, as an unconscious prehistory to much of today's art. Whitman's reputation rests on his contributions to the field of performance art, and surely the ephemeral character of such work has undermined its appreciation. But history has also largely overlooked the work of a not-for-profit group that he cofounded in 1966, together with Robert Rauschenberg and the scientist Billy Klüver and engineer Fred Waldhauer. Experiments in Art and Technology (E.A.T.) considered itself a service organization, dedicated to initiating and supporting collaborations among artists, engineers, and scientists by providing access to new technology as it developed in research institutions and industrial laboratories. Although Dia Art Foundation, in New York, initiated a traveling Whitman retrospective some years ago, accompanied by the first extensive monograph on the artist (for which I served as coeditor),[1] there is a relatively unexamined facet of his work that grew out of his E.A.T. activities, which I'll call the telecommunication projects, and that led directly to *Local Report* (2005). The telecommunications work may be seen as an early attempt, rare within the art world of the time, to grapple with the ways that nascent telecommunications technology was hinting at future modes of social relations.

Telecommunication: A Mythical and a Technological Approach

Whitman's theater pieces, together with a series of intriguing 8mm film installations, constituted his entire body of work through 1967. His milieu was an underground one, made up of small, makeshift, downtown Manhattan art spaces, often run by the artists themselves. His performances from this period are described in mythical, romantic terms, as "poetic, fugitive phantasmagoria,"[2] that seek "correspondence between nature and technology, connecting ritual and the rational, seeing computers that look like stars"[3]; dreamlike accumulations of objects and images "alchemically combined and transmuted into improbably fantastic events [imbued with] a magical, mythic aura."[4] He had gazed into the sky in his 1960 theater work *American Moon* as well as in the 1965 performance *Night Time Sky*, and, considering his later projects with telecommunications, his work may be understood in terms of these two opposed figures: looking up into the heavens and looking down from heaven to earth—metaphorically, technologically, and mythically.

Whitman created two more large-scale, collaborative works with Klüver and E.A.T.: *Children and Communication* (1970) and *Telex: Q&A* (1971), both of which may be seen, in some ways, as predecessors of such communications modes as email and Internet chat rooms. *Telex: Q&A*, which was included in Pontus Hulten's exhibition *Utopia and Visions 1871–1981* at Moderna Museet, in Stockholm, aimed at surveying what people worldwide were thinking about the future. Replies to questions concerning the world 10 years hence—that is, in 1981—were gathered from fax workstations set up in New York, Tokyo, Ahmedabad, and Stockholm.

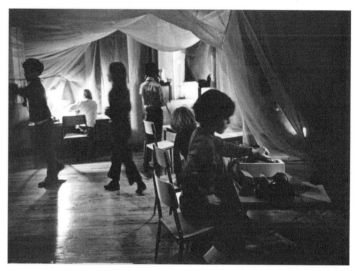

During the E.A.T. project Children and Communication, *schoolchildren work with the telecommunications equipment in the environment designed by Robert Whitman.*

Although hundreds of responses were collected, the answers have never been evaluated.

With *Children and Communication*, Whitman constructed low-ceilinged, dimly lit, tentlike environments, meant primarily for children, in two places in New York City.[5] Intended as test sites or pilot projects, the hope was that such telecommunication spaces would be set up in schools and community centers so kids could communicate with each other beyond their own neighborhoods. They were interconnected via facsimile machines, Telexes, telephones, and a machine called an Electro-writer. Telex turned out to be a popular medium for chats between children who didn't know one another, proving less abrupt than the telephone and more intuitive than the facsimile machine. Perhaps it appealed because, as with email, one was given time to consider a question before replying. Julie Martin remembered that children used the phone to teach each other how to use the equipment, and to inform the kid at the other end that a message was coming.[6] The Electro-writer also proved very popular, and it was instantaneous as well: Whatever a child wrote at one place, came out immediately at the other. In any case, all four technologies were framed in an intimate way, positioning the audience as a willing and active participant. Alexander Kluge has commented on the power of this kind of work: "One of the most effective ways of exposing the true nature of any public sphere is when it is interrupted, in a kind of alienation effect, by children. . . . In every case, the reified character of each context, its rigidity, and the fact that the public sphere is always that of adults, immediately become apparent."[7] More recently, and with similar aims, Dan Graham produced *Children's Day Care, CD-ROM, Cartoon and Computer Screen Library Project* (1998–2000), exhibited at Marian Goodman Gallery, in New York. Inside a two-way mirror pavilion, he installed computer stations equipped with games, Internet access, cartoons, and a library, all aimed exclusively at children.

Local Report

These were the largest projects Whitman undertook, but they were not his only activities in this area. In the early seventies, he assisted artists in presenting television works on New York public access television, as part of an E.A.T. project at Automation House, called *Artists and Television*. In 1972, he produced a radio work, *News*—maybe the closest to a predecessor of *Local Report*—hosted by WBAI, in New York, in which people called the station from public phones and described what they were looking at.[8] Whitman sent people to particular pay phones in Manhattan, moving from one phone to the next to report what they saw. All this was broadcast live, the montage of banal and occasionally poetic description comprised the entire radio program.

In 2002, the new media group Lumen grasped the implications of the project and invited Whitman to restage it as part of the Evolution Festival section of the International Film Festival, in Leeds, England. This can be seen as further evidence that Whitman's telecommunications projects foresaw, in various ways, the cultural shift that was to take place in the midnineties, as the growing popularity of the Internet, mobile phones,

Robert Whitman at the phone bank during his cell phone performance, Leeds, 2002.

and reality TV shows began shaping our culture. Such recent online phenomena as MySpace and YouTube continue to echo this work.

It must have been intriguing for the artist to bring a 30-year-old piece into the present cultural moment simply by switching to the latest technology. Rather than having to call from designated, stationary public phones, which had limited what was visible and what could be described, Whitman gave the Leeds participants cell phones, to walk unrestrained through certain areas of the city. Their calls were broadcast in a public square, and passersby could hear descriptions of other parts of the city as they walked or sat.[9]

Similarly, with *Local Report* Whitman was able to bring up to date both technologically and conceptually something he had begun, in 1972, with *News*. This time, he equipped people with then-new video cell phones (destined to soon become pedestrian technology) and dispatched them from five northeastern shopping centers to shoot video clips, gather individual and communal daily experience, and report on what they encountered. Whitman asked the participants to make voice calls and video calls, which came in separately but were played together during the performance. The artist made a crucial compositional decision initially to split the soundtrack from the image and spontaneously create a sound-and-video collage of these clips. Cut and sample were his methods. The separation of sound and image turns this potentially banal material into an independent, poetic, and abstract work. The image is crude, the audio hollow and tinny. Spoken descriptions were cut short whenever Whitman felt that "a coherent image was created . . . maybe because they say something wonderful, and I want to let it hang there for a while." It's striking to see how blurry, how wobbly the digital video clips are—full of jumps, blackness, silence, and images that disintegrate into their component digital elements. The aesthetic of these rather abstract images is familiar from the widely distributed images of mass-market technology: photos sent to friends or family by phone or email; most of the low-resolution images online, including Internet video; soldiers' pictures from Iraq, reproduced in the media. Whitman's piece, in

addition to entering the Guggenheim Museum's collection as a five-channel video installation, was made available online, where it remains as of this writing.

Viewing the clips online, one realizes that there is simply too much spoken narrative, too much imagery, too much data, and its transmission taxes the network, which occasionally fails altogether. As astonishing as the speed of technology's development may be, its limits mark *Local Report*. This is a paradox: While Whitman can gather far more information for his Local Reports than he did 30 years ago in *News* (when people could call only from pay phones, their voices broadcast live on the radio), the quality of today's digitized audio and visual elements is so crude that we face a work that is more abstract than ever. In this sense, the "cultural map of the area" that Whitman wanted to create is a cultural map of not only Easton (Pennsylvania), Trumbull (Connecticut), Holmdel and Burlington (New Jersey), or Kingston (New York) but, more generally, of the cultural moment we now inhabit.

Technology: The Marriage of Spectacle and Avant-garde

You might have to give up your ego sometimes because the result might not be art. —Robert Whitman

In order to understand the scope of Whitman's interests and activities in communications, it may be helpful to outline his work with E.A.T., particularly with Billy Klüver, a figure to whom he'd been increasingly drawn since the group's founding. As Lynne Cooke has suggested, their highly productive partnership effectively constituted Whitman's social and creative community for the remainder of the decade.[10] Klüver, a Swedish-born scientist who had worked at Bell Telephone Laboratories for 10 years, was once described by a museum official as "the Edison-Tesla-Steinmetz-Marconi-Leonardo da Vinci of the American avant-garde."[11] E.A.T.'s collaborative structure came out of Klüver's proclivities: "I was interested in the mind of the artist. So I was wondering what I could bring to the art world: I could only offer technology to the artists."[12] The first of these offerings resulted in Jean Tinguely's 1960 *Homage to New York*, an autodestructive machine that enacted its suicide in the sculpture garden of the Museum of Modern Art, in New York. This was followed by numerous other projects, including designing a radio transmission with wireless FM broadcast for Rauschenberg, powering a neon *R* with batteries in a Jasper Johns painting, and attaching a contact mike to Yvonne Rainer's throat so the audience could hear the modulations in her breathing as she danced, as well as supplying the material for Andy Warhol's *Silver Clouds* (1966), after he was unable to satisfy Warhol's desire for an illuminated lightbulb that floated in midair. As Klüver points out, the founding of E.A.T., which followed these early collaborations, "opened doors for artists and engineers which otherwise would have been closed."

The public climax of E.A.T.'s working career was the Pepsi Pavilion at Expo '70, in Osaka, Japan. Never before had a corporation put its image in the hands of the artistic avant-garde. Although the context was a world's fair—more precisely, a vast, public, corporate-underwritten spectacle—E.A.T. attempted to appeal not to the masses but to each individual, counteracting the authoritarian experience of the fair by encouraging participation. Entering the Pepsi Pavilion, visitors were confronted with a wondrous, spherical, aluminized Mylar mirror some 90 feet in diameter, 37 speakers that made sound throughout the space, and a floor from which preprogrammed sound emanated (audible only with a handset held to the ear), all beneath their own inverted reflections. Thus encouraged to negotiate their own experience within a multimedia environment that art critic Barbara Rose called "a secular temple of the self,"[13] they might realize that they were performers or participants, perhaps against their will.

One Global Village and 600,000 Indian Villages

The Anand Project is the most unusual among the public projects, and we can trace Whitman's key concerns with it. In the late sixties, media was not a commonplace in Asia, and most of the continent's rural population remained virtually excluded from mass communication. The decision by the United States to place the ATS-6 telecommunications satellite over India was, therefore, freighted with immense potential, since the unit could

Mirror Dome Room, Pepsi Pavilion, Expo '70, Osaka, Japan. The three-dimensional real image of the many visitors on the floor hangs upside down above their heads.

theoretically reach hundreds of thousands of villages throughout the country. This move was what facilitated E.A.T.'s unusual experiment.

Vikrim Sarabhai, a scientist who worked on cosmic-ray research and headed India's Atomic Space Program, had become interested in E.A.T. before Osaka's Expo. A cultural visionary, Sarabhai had, in 1969, invited the group to come to Ahmedabad and develop a proposal on how to generate educational programming for the U.S. telecommunications satellite.

Video Transcends Literacy

My story is about how unfamiliar the idea is to take notes, that you should make notes, locally. —Billy Klüver

The project got under way in December 1969, when Klüver, Whitman, and a team of education and technology specialists they had selected met with their Indian counterparts and commenced a series of discussions and expeditions in and around Ahmedabad. E.A.T. had printed a detailed project description in its newspaper, *Techne,* excerpted below:

> The use of television as a tool in national development is assuming high priority in India, with an increase in ground stations and a one-year experiment in satellite television planned with NASA in 1974. Virtually no experience exists in producing instructional software on the scale that will be needed. . . . The target area for the first experimental project will be the Amul Dairy Cooperative, in Kaira District, Gujarat State. The instructional objective will be to teach the care, feeding, and breeding of milk-producing buffalo to the villagers who are members of the cooperative. The objective of the project is to develop prototype procedures and facilities that can be used on a wide scale for producing television-based instructional software that will be effective in the Indian rural environment. A team of people, including an artist, with one-half-inch videotape equipment, will work in a village to compile first approxi-

> mations to the instructional material. The villagers will be encouraged to use the equipment and participate in generating ideas and materials. By using the video equipment as a stenographic tool, and feeding in reactions of the villagers immediately, the visual and aural idioms of the village are incorporated into the instructional material at the earliest "script phase."[14]

The project challenged the traditional understanding of public art and culture, especially with respect to the role of the artist. It can be assumed, for instance, that Whitman, a nonscientist with a background in performance art, was fundamentally involved in defining some of the project's groundbreaking procedures and working methods. In some ways, the breakthrough was to teach people to take and pass "notes" on their practices, a kind of self-education through the sharing of local knowledge. Instructional television in India had heretofore typically consisted of a man standing at a chalkboard writing, but new models were needed in the largely illiterate rural areas of the cooperative, which is where the videotape was crucial.

The material for the Anand Project was to be generated with the involvement of the local people, who made videotapes of women milking, artificial insemination, the treatment of sick buffalo, and so on. This video

Buffalo being led to shed for artificial insemination procedure at the Amul Dairy Cooperative.

chain letter was distributed from village to village by the dairy delivery boy, and people from different areas added their own videotaped comments on what they saw; the finished tapes were brought back to a studio in Delhi and used as both the visual and structural basis for a professional TV program. They had a dual function, then: Far before their role as viewers of the broadcast, many of the villagers were already direct participants in the making of the program material. According to Klüver, that way the studio culture wouldn't be imposed on the local people, "because you start from their situation."[15] This active involvement echoes the notion of transforming all spectators into participants, an idea present in many sixties performances, notably in the work of artist Allan Kaprow, with whom Whitman had studied. Different from his teacher, who gave participants in his pieces precise instructions on what to do, Whitman invited visitors in the Pepsi Pavilion to explore the environment on their own. He created situations that entrusted the audience to create its own content, as with his later *News* and *Local Report* pieces. They also operate with the idea that the work will be made by the local people who participate. Whitman preferred to give up the authoritarian control that he saw in the BBC television studio aesthetics or in the productions of conventional theater directors.

The Indian SITE program was put into effect some years later, in August 1975, when the satellite began transmitting to "some 2,400 villages in six states . . . [and rebroadcast] to another set of approximately the same number of villages."[16] Each village was taught how to fashion a parabolic antenna from chicken mesh, and electricity was generated by bicycle power. One television was distributed to each village by the Indian government, and a later study estimated that there were 80 to 100 viewers for each set. An estimated five million people viewed 1,200 hours of programming in one year, and the satellite had the potential to reach all of the country's approximately 600,000 villages.

India was the first country in the world to use satellites to transmit educational television programming directly to its citizens. According to science-fiction writer Arthur C. Clarke, who, in 1945, was the first to propose the concept of a geostationary communications satellite, this "generated new capabilities, demystified space technology, and helped to nucleate a large island of self-confidence."[17]

The satellite program lasted only a year, and some years later the Indian government lost its television monopoly, making way for more diverse and competitive programming. As with most mass media, this meant profit-oriented commercial entertainment.[18] Nevertheless, the concepts and ideas of the Anand Project proved a potent window of opportunity to improve the lives of hundreds of thousands of villagers; it stands as an example even now. In his 1970 project notes, Klüver observed: "The experience gained from this experimental situation would be directly applicable to the wider context and larger audiences of ground or satellite transmission of television."[19]

In retrospect, it may seem that excitement about the utopian aspects of the new technology was uncritically optimistic, but it was typical for the sixties. The hope that these tools would yield a truly democratic public culture were most famously expressed in the writings of Marshall McLuhan, but the era's idealism can even be detected in Henry Kissinger, who was involved in the decision to place the U.S. satellite over India:

Satellite technology offers enormous promise as an instrument for development. Remote sens-

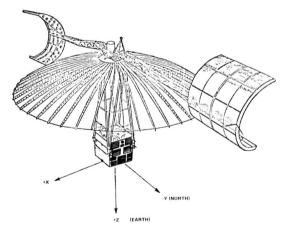

Diagram of ATS-6 communications satellite.

ing satellites can be applied to survey resources, forecast crops, and improve land use in developing countries. They can help foresee and evaluate natural disasters. Modern communication technologies, including satellites, have large untapped potential to improve education, training, health services, food production, and other activities essential for development.[20]

Seen within the context of the Cold War and the space race, these words may seem steeped in ideology, but even today we must acknowledge the dark side of communications technology, which can be used for aims of terrorism, nationalistic propaganda, and corporate exploitation and control. Osama bin Laden and many other fundamentalist copycats have repeatedly employed the video letter, and the Chechen rebels who stormed the musical theater in Moscow were producing a similar broadcast (with the forced cooperation of the musical's director), but were killed by police before it could be completed.

The Artist As Amateur and Observer

The artist carries the least cultural baggage.
—Robert Whitman

Whitman was an obvious choice to participate in the Anand Project, not only because of his close friendship with Klüver and his work with E.A.T. but also his self-acknowledged openness, curiosity, and anti-imperialist approach in involving local populations. His notion that "the artist carries the least cultural baggage" was possibly even more applicable outside his own artists' culture, where his outsider status allowed him to "see wonderful things in anything that others might overlook."[21] Klüver has made a more historical suggestion regarding Whitman's role in the project: "In a complicated situation, where decisions have to be made about software and hardware, the artist is the best collaborator. During the 16th century and earlier, artists were always called for when complicated decisions had to be made in large projects like ship-building and cathedral building."[22]

As with his later *Local Report*, Whitman's central role as producer in his own artwork was counteracted by the project's challenge to traditional authorship, in the move of farming out its content to many people. This was fine, since his real goal was to "push the art into the social." In this light, considering its scale and ambition, the Anand Project must be seen in the tradition of those historical avant-gardes—most famously, the Russian avant-garde—who attempted, through art, to effect social change and alter the experience of daily life. In this case, the key to this ambition was the total lack of mass media in rural India, which practically demanded the utopian challenge to initiate a program for an entire *populus*. As with the Russian avant-garde, it was a project outlined by a small elite, who saw themselves entrusted with the task of developing a culture from scratch.

Today, it is not uncommon for artists to work with new technologies, and many engage in socially active practice and collaborative work arrangements, with an anti-imperialist view founded in social science and critical theory. (Consider, for instance, the Delhi-based Raqs media collaborative, which was invited to the 2002 Documenta, or the New York–based group 16Beaver.) This may be read, perhaps, as the response to a global, corporate-driven culture, and also as an outgrowth of academic work on postcolonialism, and the broader dialogue on multiculturalism. In 1969, however, the Anand Project was decades ahead of its time. While Frantz Fanon had made a stir with his 1952 book *Black Skin, White Masks*, postcolonial theory didn't exist as an academic discourse in the late sixties, and Edward Said, author of the landmark text *Orientalism* (1978), hadn't yet published at all. Furthermore, the project's technology was absolutely cutting edge: Video was not to enter the art world, let alone the consumer marketplace, for several years, and satellite television was itself a major innovation.

These novelties, however, enabled the project's truly radical aspects: It was not simply about new technology but about finding a viable solution to the problem of how to introduce this technology to developing countries. With the Anand model, the assumption was that one could minimize the effect of first-world cultural biases on third-world recipient countries. E.A.T.'s

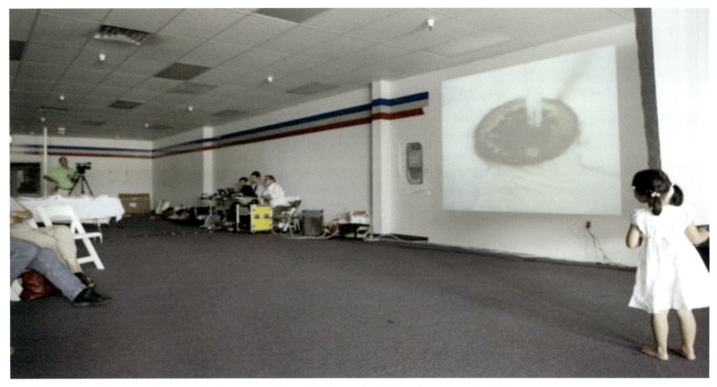

Burlington, New Jersey, performance headquarters, Robert Whitman, in white shirt, at control table.

general concept and infrastructure were adopted by the Indian government (even if only for a year), which turned them into what Klüver called "real life."

Not Pop but *Populus*

In his collaboration with E.A.T., Whitman proved a stubborn participant, never quite able to leave his underground art world behind. "I haven't called the scientists, I hate to take up their time," he said, and: "I want to go back to ropes and pulleys and yelling for communication."[23] His career is marked by ambivalence. On the one hand, Whitman embraces new technologies and the idea of an audience as active participant, particularly in his telephone pieces; on the other, his own work from the sixties couldn't be more different from the project in India. In 1974, he temporarily withdrew from E.A.T.'s public mode of working, returning to his studio, where he worked on the *Dante Drawings*, a series of 27 double-sided drawings based on Dante's *Paradiso*, a text

he had studied closely in the midfifties. This was a time of introspection for him, and it affirmed his "trust in drawing as an activity of radical opposition."[24]

Yet, from early on, Whitman had pondered the issue of "the blurring of art and life," the words of Kaprow that he would reinterpret throughout his career. Many others were similarly concerned with this problem, including friends and collaborators involved in his early theater works, such as Claes Oldenburg, Jim Dine, Rauschenberg, and Red Grooms. These artists became key figures in the pop movement, trawling the ocean of mass culture and everyday phenomena, elevating what they found to the realm of art. Their museum work reflected on and critiqued the aesthetics of a relatively young capitalist society driven by consumption, mass media, and popular culture, and in this way forever altered the syntax of art. With his telecommunications projects, however, Whitman took the opposite tack, reaching down into the *populus* from above, "escaping the isolation of the art world, into a functioning relationship to the larger community"—or, as Klüver once put it, "working realistically in a society."[25]

Notes

An early draft of this essay appeared in *Printed Project* 1, Sarah Pierce, ed. (Dublin: Sculpture Center, 2003).

1   Lynne Cooke, Karen Kelly, and Bettina Funcke, eds., *Robert Whitman: Playback* (New York: Dia Art Foundation, 2003).
2   Cooke, "Robert Whitman: Playback," *Whitman: Playback,* 12.
3   George Baker, "The Anti-Images of Robert Whitman: *The Dante Drawings," Whitman: Playback,* 120.
4   Sally Banes, "Dream in a Warehouse," *The Village Voice,* November 23, 1982, 105.
5   One of the locations was at 9 East 16th Street, in Manhattan, and the other at the Automation House, 49 East 68th Street.
6   Email from Julie Martin to author, January 18, 2007.
7   Oskar Negt and Alexander Kluge, "The Public Sphere of Children," in *Public Sphere and Experience: Toward an Analysis of the Bourgeois and Proletarian Public Sphere* (Minneapolis: University of Minnesota Press, 1993), 283; originally published in Germany, 1972.
8   *News* continued to be presented over a two- or three-year period in various cities in the United States, including Houston, Minneapolis-St. Paul, and New Brunswick, New Jersey.
9   In 2003, Whitman proposed to make the piece in Tokyo at the time of a large E.A.T. exhibition, in which he wanted to use the most recent cell phones that could also take pictures, but it couldn't be arranged in time.
10 See Cooke, "Whitman," *Whitman: Playback,* 16.
11 See Calvin Tomkins, "Onward and Upward with the Arts," in *The New Yorker,* October 3, 1970, 86.
12 Billy Klüver, interview by the author, May 12, 2003.
13 Barbara Rose, "Art As Experience, Environment, Process," in *Pavilion* (New York: E. P. Dutton, 1972), 99.
14 "The Anand Project," *Techne* 1, no. 2 (November 6, 1970), 7.
15 Billy Klüver, project notes, March 2002; Klüver/E.A.T. Archive.
16 Ernest Corea, "Rich Men's Toys or Development Tools," 19; Klüver/E.A.T. Archive.
17 Sir Arthur C. Clarke, "Satellites and Saris: 25 Years Later," *Frontline* 18, no. 9 (April 28–May 11, 2001) http:/www.hindu onnet.com/fline/fl1809/18090100.htm.
18 According to an evaluation of the Satellite Instructional Television Experiment, the audience preferred instructional programs to pure entertainment.
19 Klüver, project notes, 1970; Klüver/E.A.T. Archive.
20 Henry Kissinger, quoted by Corea, "Rich Men's Toys," 19.
21 Whitman, interview by the author, May 8, 2003.
22 Klüver, project notes, May 2003; Klüver/E.A.T. Archive.
23 Whitman, quoted in Simone [Forti] Whitman, "Theater and Engineering: An Experiment. 1. Notes from a Participant," *Artforum* 5, no. 6 (February 1966), 27.
24 Baker, "The Anti-Images," *Whitman: Playback,* 114.
25 Cooke, quoting 1977 notes from Klüver, "Whitman," *Whitman: Playback,* 28.

Local Reporters watching the video and sound replays at Northampton Crossings, Easton, Pennsylvania.

Local Reporters watching the video and sound replays at Kingston Center, Kingston, New York.

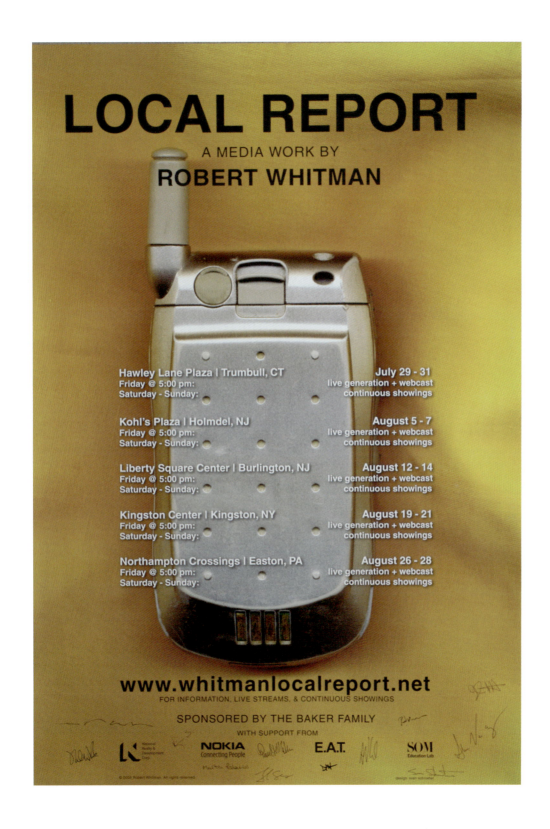

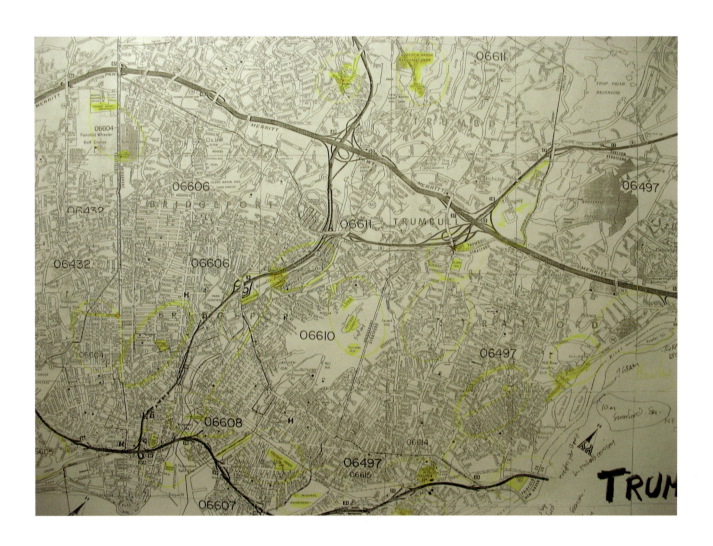

The park by Captain's Cove is really run down; I would not like to be a little kid and grow up and play. There are kids running around, a few people motorcycling over grass. • One airplane is being taxied out from the airport—that's where I am—while another small one is being pulled in. It's a beautiful windy day out here, what with the airplanes, broken glass, blacktop. • We're here at the local bar talking to the pilots, who are all sitting around drinking, having a great time. There is a dog in the doorway watching a fly outside. There's a lovely lady behind the bar. And it's a warm sunny day, and it seems to be a lot cooler here inside; apparently an old farmhouse, that's been . . . • We're at the playground on Saint Stephen's Street near Captain's Cove, and across the street is a park, which is real filthy—daunting. This playground, though, is clean and pretty; and only a couple of kids have just shown up. Otherwise, it's really empty, which is also kind of an oddity [unintelligible]. There are a lot of people

out on the street. The air is beautiful and warm, and a sea breeze, and it's quite comfortable. There is now a family playing on the playground. • Sacred Heart University, Fairfield, Connecticut. I can see the William H. Pitt Health & Recreation Center, and I can see the Campus Field, "Home of the Pioneers." Very blue sky, lots of buildings. • I am standing here in front of the Children's Museum; it's actually the Children's Garbage Museum. And I'm facing to the left now, looking at these gigantic silos with chalk drawings, by some very industrial area. And then you have this very cute museum, with a small building and a little sculpture out front that's made of cement, and some glass, and a green door, and a spiral staircase going high. And it's all very charming, and it's attached to . . . • I'm at Saint Michael's Cemetery, and I'm looking at a tombstone that's tilting. The name is Mary Carroll, died in 1889 on December 27. The design looks like it's been just washed away with all the rain and the weather.

It's kind of a rosy-red granite. She died at age forty-eight years old. There are leaves visible on the bottom of the tombstone. • Hi. We're up on the hill on Madison Avenue, in Bridgeport, right next to the baseball field, listening to the wind in the trees and the birds. We can see seagulls, and also across the street from the Correctional Institute. • Hello. I'm at the harborside down by the docks, and we're relaxing in the lovely town of Bedford. We're about to order another drink; and the people here have been very welcoming and hospitable. And they said we could photograph them and use the phone and the videophone here. • [Background voices] (Say it again. Is it a talking one? Yeah.) Oh, all right, here we are at a busy corner, lots of cars and trucks going by, and people going in and out of shops. We have people pulling up to go, probably, into this wonderful bakery with twenty million calories in it. And, then, down the road there's two gentlemen building a new storefront. • I'm in a parking lot.

There's a fair amount of traffic for this time of day. There's an industrial complex across the way that seems to be available. I wonder if the entire thing is available. [Unintelligible]. I'm walking toward a line of trees. • We're high on a hill above Bridgeport, on Madison Avenue, across from the baseball field, listening to the wind in the trees and the birds. We can see seagulls and the Correctional Institute. • We're reporting from the intersection of Church Hill Road, Daniels Farm Road, and White Plains Road. Across the street, there is a Mobil station, and gas is two forty-seven a gallon for regular, and people are filling up like crazy. Everything else seems sort of like a normal Friday traffic day. Everyone zooming by, no one is speaking. • The zoo is closed as closed can be; it's been closed since four o'clock, so off to the next thing. Bye. • I'm standing here looking at the trees. I see an old grave by an old, old house, and right near it I see some beautiful, beautiful old oak vines going up the trees.

• O.K. I'm starting. (Yeah?) Well, we're at Frash Pond. We're at Frash Pond—(Frash Pond, yeah?) Yeah, we're at Frash Pond. Unending stream of cars [unintelligible]. There is a big construction firm here, a big-sized . . . • Hey there! Calling from Short Beach; watching a little tabernacle, a family picnic going on. There's about thirty people; they're all eating lots of hot dogs; kids playing, and lots of smiles everywhere. A beautiful afternoon and the sun is shining; see sailboats in the distance. • We're washing our hands, and people are leaving. Nobody knows that we're really videoing things and talking. I feel incognito and sort of like a spy. Over and out. • [Unintelligible Spanish] • Saint Michael's Cemetery: We're looking at a tombstone that has fallen down; it looks like somebody pulled it down. There's a big webbed strap around it, and there's a big broken bar beneath it. The date says 1844, and the person died 1903. It appears to be, maybe, pulled down by somebody; maybe it was

vandalism. I'm not sure. It's a gray stone with specks, glittering specks of mica or something in it—maybe quartz. • I'm sitting on the beach. Kids are playing—sad parents with the kids are playing. It's breezy and lovely. Birds are making noises. Whoa! [Bird noises] • [Unintelligible] • We are just off of Route 8, going to the . . . toward the Beardsley Zoological Garden and the Wonderland Ice Park, Bennett's Pond. Grinnell Pond is there, with a waterfall. And some kids are coming out with a hockey stick in their hands. There is, ah, several intersections with lots of cars, and it looks like the sun is bothering everybody. Father driving. And there is a Rent-a-Center, a Goodwill. • Ah, yes, I'm at the [unintelligible] yard in Union Street. [Unintelligible] There is some kind of equipment and some kind of warehouse. There was a truck coming out of the warehouse. And, basically, it's not much . . . • Hi. I'm out here down by the dock, and it's a beautiful day. There's a soft breeze blowing; it's not too hot and

it's not too cold. I feel very contented. I'm enjoying looking at this big Harley-Davidson. • I'm at Short Beach, in Stratford, across from Sikorsky Airport. I see children on the beach playing—a woman, probably their mother, in a blue bathing suit. Children in the playground, a mother and child . . . • (Well, anyhow, I'll tell you what—you can sit right here.) O.K., I'm in Stratford at the Duckpin Lounge, surrounded by a lot of gentlemen who are [men talking in background] sitting down to my left here. And I'm eating peanuts. • Yes, I'm in Stratford. I'm looking at an old, early American cathedral, all white, and next to it are some tall, tall elm trees. Everything is all green. • (Play that again.) Hello. Numbers twenty-one on the Honey Bus, sha-boom, sha-boom; Honey Bus, sha-boom, sha-boom. • I'm in church, and there's a little day-care center, and I guess this is the kind of—I don't know what they call it—the minister's house, next to the church, with the junction of Route 115 and Route 130. A couple of guys

 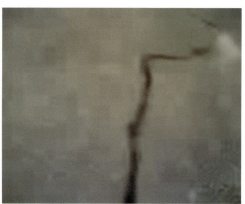 

 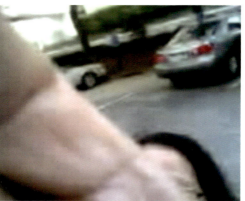 

who look like they are fixing a porch, or taking it apart, are taking a cigarette break. And apart from the guy who's looking for his [unintelligible], there's nobody walking around here. And it's very strange and very safe. Lawns: very picturesque, if you ignore cars, O.K., and the concrete. O.K., thank you. • It's like ice. Sitting on this bench, watching these people down here at the pond outside the closed zoo. It's such an idyllic day—too easy. This was the best assignment ever. • I'm staring at a cigarette butt on the ground, in the leaves with sand. I'm out here on Long Island Sound, on the beach, looking at boats that are watching people watching boats. The wind's blowing in the volley-ball net—and they need to fix it; actually, it's torn. Oops, there go the birds; they're flying away. • Saint Michael's Cemetery—looking at a huge tombstone that has an arch that you could walk through. And there's a white marble monk holding a baby. Interesting thing is that on each side there are actual photographs

on oval-shaped plastic. • I'm standing here in Stratford, looking out on the water and the sailboats going by, near the Stratford Paper . . . • I'm in Stratford, Connecticut. I'm looking at lots of people having a barbecue underneath a sort of awning thing. They're eating hot food, and it looks really good. They're having hamburgers, and I'm going to go . . . • (I'm not very good at . . . I'm not very good with technology. My video report failed to go through. I don't know why.) • I'm standing on the first floor [sound of motorcycle] [unintelligible] . . . which is old Stratford, Connecticut, facade on a facade on a facade. And the smells coming in the open window on the sidewalk are incredible. Gonna go eat. • This is from Indian Head Park, in Trumbull. The children's play area, which has sprinklers for the children to run through. • I'm standing on a green lawn in Beardsley Park, overlooking Bunnell's Pond. There's some shade trees on my right. In the distance, a large flock of Canada geese, and a

mother kicking a ball with a few toddlers. • (There's just static, does that mean . . . ah.) That's the train station. In a way, the train is kind of the least present thing here. We've got cars—I-95. The whiz of trucks and cars are the dominant presence. Even this train station is not an actual station house; instead, it's the National Helicopter Museum. • [Beep, then silence] • Still at the beach. (I think the music has stopped; I'm not really sure anymore.) A couple of birds are sleeping, their heads tucked back between their wings, their legs tucked under. Oh, one just got up; he's moving around, I think he's [unintelligible]. So, it's people, beach stuff. (I really do have to pee . . .) • [Sound of traffic, static] • The 5:35 commuter train from New York, the 5:25 Port Jeff Ferry, 1:18 bus from Stratford, "Ferry Parking Next Left." Newspapers swirling in the wind. • (No, I don't think it even re-corded it. [Voices] Some guy said something. [Voices] The music was stopped.) • So, we're sitting here, and it's warm and it's

humid. And I wonder who came up with the idea of a heat index. It feels like it is ninety-five. • Hi, I can't tell whether the music has stopped, but I'm hearing myself talk, so I'm going to be optimistic. I saw a front yard in the middle of Greenfield, Connecticut, that tried to have a desert with gravel and rocks, but they had no palm trees. I saw three geese trying to find afternoon snacks in the lawn next door. • This is from the amphitheater at Indian Ledge Park, in Trumbull. The amphitheater is used for many different functions. Last year, it was used for a sell-out crowd of the Beach Boys. This year, it had a three-day Irish Festival that was held here for the first time. • Hi, I'm in the [unintelligible]. I'm in Bridgeport Hospital, in Bridgeport. And it's in a very nice building, but there's nobody here. Cars are going in the main street. And it's a white building with very modern—what do you call—windows, and there's some . . . • Yeah, real life always interferes. Past and present simultaneously. I'm concentrating

 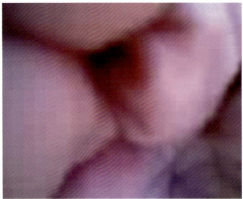 

on what I'm doing, and then I get a phone call, and I look to see the image I'm videoing—video? Yeah, and it transports me back in time thirty-five years, and then I'm on the phone, and I'm in the present, and it's not. I'm not even thinking about . . . • Here we are in front of Target. We're looking at where Best Buy will soon be over at the Hawley Lane Mall. There's a lot of customers here, going in and out, and a lot of people that work in the shops are collecting carts, and customers are leaving. • Here at Route 3, at the *Local Report,* and we are enjoying a gorgeous afternoon and the water. And we can see the traffic going along, and we are under a big . . . two gorgeous trees. It's a little bit cloudy, but we still see the sky. • O.K. Hi. Down by the river in Stratford, looking across at condominiums that look like pigeon coops. I saw two men on lawn mowers that you can ride on. And they were trying to talk to each other, and one of them had headphones on to protect his hearing. • I'm facing west

again, on Post Road, in Stratford. I'm looking at a sign, "Elms at Stratford Plaza," C-town, with all kinds of stores, including [unintelligible] store. • Hi. I'm outside the back of the Hawley Mall, on the upper level. I've just seen a couple come out with a very pregnant woman who's expecting a child. It's the second pregnant woman I've seen in about two minutes. And that's just about everybody going in and out, it seems. • On the corner of Main and [unintelligible] Road, walking; on the left is a chain-link fence with green slats, rubble, sidewalk cracked, blue gas line, manhole cover . . . • I'm walking by the abandoned concession stand in Beardsley Park; the dark green building with red trim, it's hung with some white Christmas lights, in no particular pattern. It's sort of hung over a sign that has no lettering. In front of it, there's a bunch of Lazy Susans, black . . . • We're at Bunnell's Pond, and we're trying to get the Canada geese to make noises. The water is calm. There's people playing with their dogs

in the park, children and their mothers playing with balls, running down the slope. There's beautiful . . . a set of beautiful trees: Copper Beech. • Beautiful as a butterfly [child's voice] . . . • Oh, Route 111, 127; I'm always interested in the strange juxtaposition. I'm wondering what that sign's doing there and what it has to do with this location. • We are leaving the Honey Express Road, going down [unintelligible]. We are seeing a gas station, very colorful. • [Unintelligible Spanish] • (Oh, yeah, O.K., don't tell me, but I'm standing in line.) Well, I'm looking at a lot of people entering the Marriott, [unintelligible] perhaps beautiful flowers. Looks like a whole busload of people are just arriving. See children drinking soda. • I'm at the Bridgeport Hospital. I'm in another building; it's more activity here. And there's a couple of people waiting in front of the doors. And I can see many, many units. And they have another one of those connecting things—one building and another one. You can see the city down the hill.

I'm in front on the avenue, and I guess it's the hospital. • This is from the soccer field in a park—Trumbull. The soccer field was recently re-covered with artificial grass to make it easier for the players. To the left is the red building, which is the teen center. • We just drove past the road where—that's the Hawley Lane Mall—there are beautiful trees. Opposite it is a carved-out area of market, and the contrast is really striking. Most of the trees are very young, beautifully twisted, and completely the opposite of the big round balls of Target, which are nice in their own way, but . . . • [Sound of seagulls] Peanuts and popcorn and Cracker Jack, burgers, hot dogs, Red Hots. Mmm, smells good. Seagulls are driving me crazy out here. But the wind is nice—a cool breeze in the shade; otherwise, it's very hot. Hotter than a match. • Park scene: All the fine young men getting buff. • There's lots of cars going in both directions. A bird just flew in front of my car. • I'm standing at the entrance of the Beardsley

Park Zoo, where people are beginning to gather for the Shakespeare in the Park tonight. They've got their coolers and blue chairs. There's a greenhouse in the distance, colorful flags waving in the breeze. • Yes, I'm at the [unintelligible], and I think there are people resting. Taking pictures from my car window, and there was some person pointing out the car. Some people were sitting on a bench, talking. And there's a little more activity of people around. (And right now, I will be taking more reports because I can see people around, but I don't know if you're supposed to take people or buildings. But I'm gonna try to take another one, and then I'm gonna quit. I'm heading back after this next one that I'm taking. I'm going to concentrate on one door. All right, yeah, I'm waiting for the next report. Am I through?) I'm still at the hospital, and, like I say, I'm going to be reporting in the front door. This is more than forty seconds, so I think I'm gonna hang up.

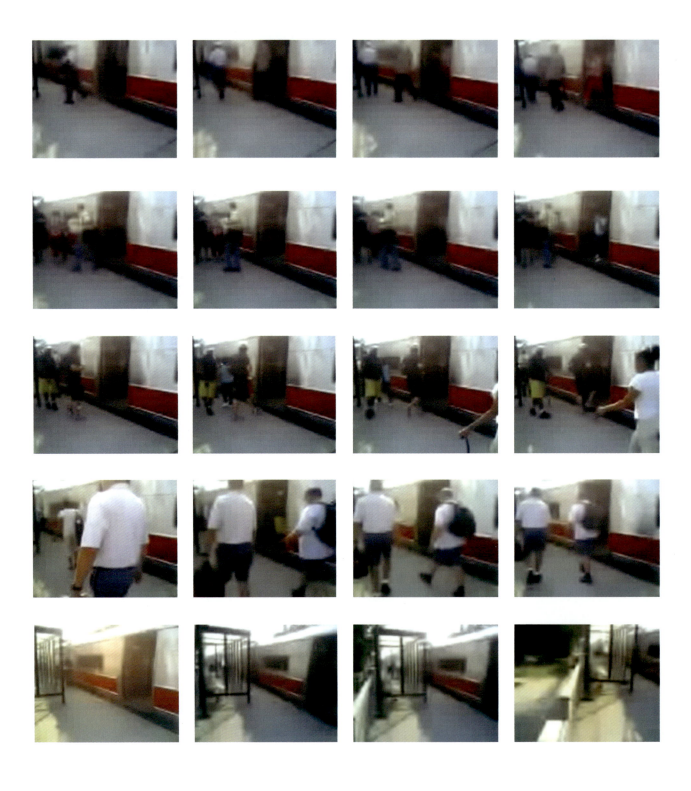

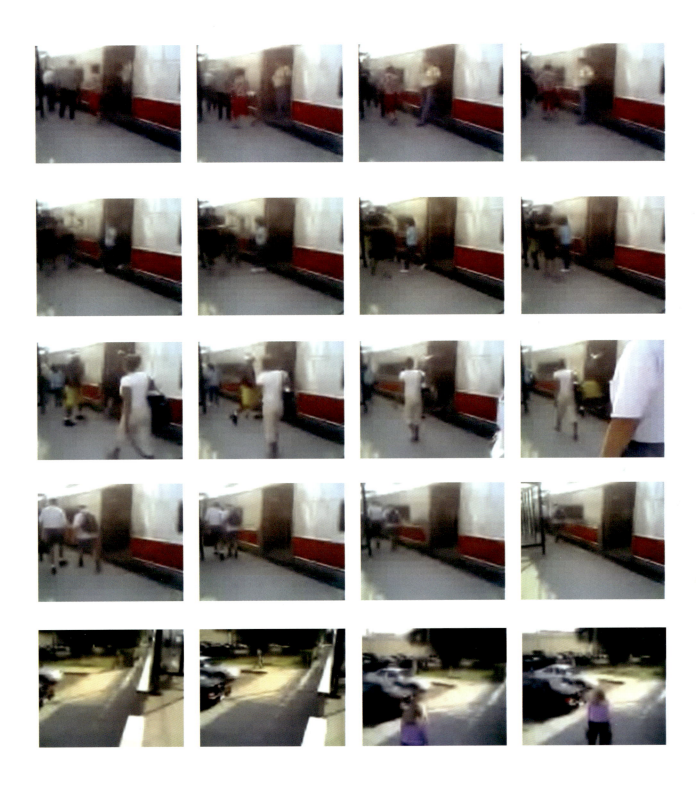

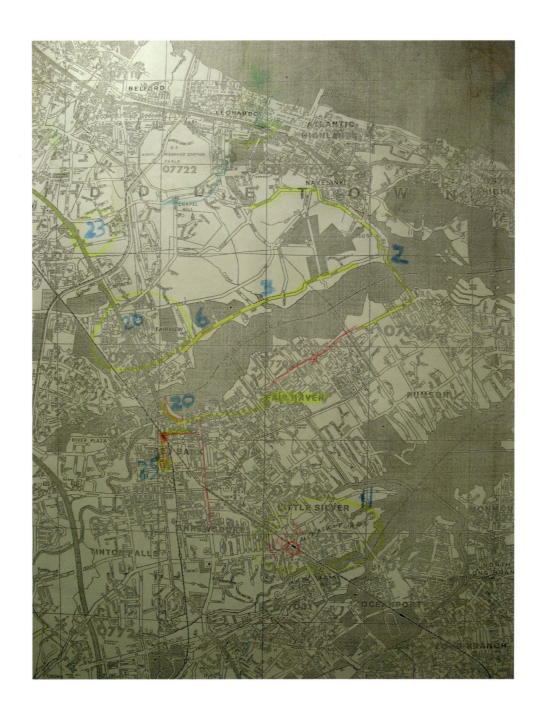

Horn Antenna has been designated a national historic landmark. This site possesses national significance in commemorating the history of the United States of America. Scientists Arno Penzias and Bob Wilson—with the antenna—found the evidence confirming the big bang theory of the creation of the universe, forever changing the science of cosmology. • Hi. I'm recording here, right in Red Bank, on Bridge Avenue. There are several people that just got off the train, which may— Well, I'm not accustomed to looking [at] what other people in this area . . . they've got. I won't go into all of the details, but, to be repetitive, quite—I don't know—different. Train just went by. Actually, it was there for a while; not sure if there was a problem or something. There are a bunch of cars. There's a lot of Mexicans. If you know Red Bank now, lot of . . . • Hi. I'm in Little Silver train station. I am currently standing outside of a nail salon, and there is a woman getting her nails done. It looks like a local

commuter train is coming in as I speak [sound of train]. Very loud. • Hi. I'm standing in a butterfly garden at Deep Cut Gardens, in Holmdel. Five o'clock's perfect time of day. There are hundreds of small white butterflies, and a very large yellow butterfly fluttering around. One of the park rangers just got soaked trying to replace her sprinkler. That was very amusing to watch. • I'm standing on the outskirts of Tatum Park, Red Hill Road, the very same road that General Cornwallis retreated down on the heels of George Washington's army. Before me, all shadows blocked out. Before me stands a massive oak tree. I'm sure this oak tree witnessed that retreat. I know that tree is growing. • Hi. I'm in Middletown, on Church Street, outside of Christ Church Episcopal Church, more than three hundred years old, founded in 1702. Signing off. • The gates have risen. The trains have left. The flags are flying in the wind. An old man has locked up his bike. The train that he came on,

taken, is gone. The roads are full of people who are itching to get back to their homes after a long day of work; high anticipation for the weekend. • I'm standing at the corner of Route 36 and Poole Avenue. I see a telephone pole with flowers attached to it; above that is a picture of someone who, I am assuming, lost his life at this intersection in an accident. Very sad to see this. That's a very sorry corner. • This is Amy. I'm in Belford. I see ships out on Raritan Bay. I see an American flag blowing in the very strong breeze. I see the Spy House. I see a deli across the street. I see a wooden fence. I see cars in the parking lot. I see my daughter fixing her hair in the reflection of a car window. • Hey, we're in beautiful downtown Keyport, and we just stopped in First Wok, got some egg rolls, and we're at the corner of East Front and Broad Street, and we're going to head down to the water. Quite a bit windy here, lot of activity. I can see a little baby. I see people on cell phones, a

lot of cars on the street, a truck. • This light is not meant for Plexiglas. It's frigging tape, hard hat, air compressors, paint trays, two lengths of wire, big things that hold signs, a work stop all lit up. • I'm in Middletown, on Middletown-Lincroft Road, just sitting at the train crossing, at the level crossing. The train just went by. Traffic is just starting up again. People getting off the train; they look really hot. • Hello. O.K., I'm driving. I see a large parking lot filled with cars and a train station. There's not much around. • I'm standing on Red Hill Road, in Holmdel, near Tatum Park, at a location marked to be an area where Cornwallis withdrew his British troops on June 29th, 1778. Not much action here except a few SUVs driving by. I guess it's [unintelligible] work for the day. • I'm standing in some marshland with my feet in the mud. I feel like I am in Africa, and looking at the little river with a few boats on it that are people's backyards. There are high grass and lots of fallen logs. • Hello!

Very windy down by the water. The storm may be blowing in, ah, probably be in about an hour. And lots of nice boats, and John's fishing boat—Captain John's. • I'm standing in W. H. Potter and Sons Lawn & Garden Equipment, facing the mouth of the tributary off the Garden State's very own [now] river: the Parkway we call exit. • Cars swimming, swimming home to do whatever they do until. In back of me are big, empty fields, empty with the exception of some boats. • I am in Belford. I see some cars. I see the ocean. I see a flag. I see some trees. I see Jersey Joe's, and I see the Spy House. • I am reporting live from the Raritan Bay, in Belford, New Jersey. The wind is blowing about five to ten miles an hour. People are crabbing, and wildflowers are blooming on the sand dunes. • There is a man sitting, encased in a phone booth. There's a blue *Asbury Park Press* newspaper vendor. • I'm in Little Silver across from the Health Fair. There is a red pickup, and somebody is

stealing the trash here. They're loading up their pickup with some old lumber and some old pipes, and they seem to be getting some more as I speak. And now they're . . . They seem to have enough so they're taking off. They've left enough lumber in there. And now they're on their way, turning around . . . • Hello. We're calling from the Atlantic Highlands Pier. We are walking along and seeing all these great boats and watching a family of fishermen getting ready to go out. They are baiting their hooks and loading their boat up, and they're about to take off. It was nice light, and we're waiting for the commuter boat to come in and bring the ferry. • There is a bright orange "Yield To Pedestrian" sign in front of me. A gray car with a New York license plate. I see fuzzy dice on the dashboard of the car. • Keyport Historical Society [unintelligible] boat dock. Keyport On Wheels Soap Box Derby, Battle of the Bands, New Jersey Coastal Heritage Trail. • (Am I on?) I'm on the Oceanic Bridge,

over the Navasink River. I'm looking down at the river, and a couple of the Jet Skis are going by. I'm watching the cars come over the bridge, and there are no people, just cars. • Hi. I'm at Bayshore Waterfront Park. I see a Dodge Grand Caravan. I see trees, I see water, I see a jetty. I see Monmouth County Park vehicles. I see boats are in storage up in large metal frames. • I'm leaving the old part of Middletown, heading back out to the highway, where I see old homes, and back to the strip malls. • Yes, I'm at Lucent Technologies, in front of the giant transistor-shaped water tower. Not much going on—car trickling out about every ten minutes or so. I guess these engineers do like their overtime. Not much else to report. I don't think it's a water tower, must be something [unintelligible]. • Hi. I'm at Debora Farms, intersection of Kohl's Plaza. T. J. Maxx is here and so is the Whitman *Local Report.* I'm reporting to say Anne-Olivia is here with me now, and she is on her way to a 7-11.

This is what I have to report. It's very peaceful here, and I like it a lot. • I am standing in a shaded path in the middle of the woods. There is nothing but a lot of overgrowth, and I have the feeling of being in the middle of nowhere. • I am on the waterfront of Lake Lefferts, looking at houses across the lake, looking at the forest and watching the birds go by. • I am reporting from the parking lot of the Shiki's Japanese restaurant, on Highway 35, in Holmdel. Two men are in the parking lot putting up a new sign that says, "Check out our new décor." • I am in Middletown, outside of the Whole Foods store, in the Chapel Hill Center, and I'm sitting on [unintelligible] organic compost. I'm sitting on a bag of compost. • Originally, Keyport Plantation was settled in 1717. (By the who?) By the Kearney family. The town began its growth in the early 1830s when the plantation was divided into building lots. • I see a "Do Not Block," a "No Alcoholic Beverages" printed sign. I see a lot of plants, and I

see a garbage can, and I see a whole bunch of piles of rocks, and I see not too many cars. • Hi! I'm here in Little Silver. I walked down a side street, a dead end. I've met a very nice family and their dog. I'm currently talking and telling them all about the project. They've asked me about whether I had some I.D., and I told them I'm a local businessman, so hopefully . . . • I see a water tower in the distance. There are several seagulls, and the sky is overcast. I see my boyfriend standing on a large pile of gravel. I see a port-a-john. I see more trees. I see a dirt road. I see more seagulls. There is land in the distance. I see water. I see my car. I see my daughter. I see more . . . • I'm on the Oceanic Bridge, still looking out at the Navasink River by the channel markers. There is a sand spit with a lot of birds on it, but I can't tell what they are, and below me, if you look right down in the water, there are a lot of jellyfish going by. There is one fisher boat out. • I'm at Chapel Hill

 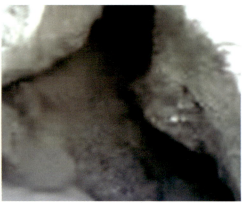 

Center, in Middletown. There's a girl trying to fit her very large bike into the trunk of her SUV, and she is having a lot of difficulty. There are two rows of shopping carts there, perfectly lined up; parking lot is full of cars and . . . • I'm standing at the corner of Lake Bridge. I am watching a man sit, while looking at a boat rack. • (Shhh . . .) Hello? Hi. I'm currently in Middletown, New Jersey, right now. I'm now on Route 35, and I'm at the Whole Foods store. There are a bunch of people out here, and there's actually a woman who's [unintelligible] with a bike, and trying to get it out of her car or in her car. I'm not sure which one it is. Also, a bunch of people schmoozing in the handicapped spot who—now they look like they are going to [unintelligible]. That might be a little abusive, yeah. • I see the numbers *4, 7, 2,* and the letter *B* on a light post, and it's section 1E. I see a Mountaineer truck. I see a "Yield to Pedestrian" sign. I see a Mets sticker. I see a duff hat, and I see an FX4 off-road

truck. • I'm looking across the street, next to Lake Lefferts, onto a house that has a van on the front lawn, with a shed. The house is pink with a pink shed. • Standing between two tables over an empty garbage can—in front of a brick wall on a white sidewalk, with an arrow facing towards Main, also saying, "Do Not Enter." • I'm in Colt's Neck, at Highway 34, at Delicious Orchards between 537, and I don't know the name of the other road. There's a light breeze in the air. • I just drove over Pews Creek. I saw a Domino's Pizza driver go past. There's a couple of swarms of blackbirds. I'm on Port Monmouth Road. I see a Little Tykes basketball hoop, I see a boat that's for sale, I see Phragmites, I see a gravel road, I see yellow . . . • I see a sign, and it's a train station sign, that says, "Track 1 to Bay Head." I see a blue dumpster that says, "Newspaper Only." There's also a newspaper sign that says, "*USA Today*." I see three, four bags of trash. • "For Sale" signs, boats on trailers,

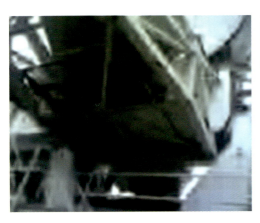  

houses in disrepair, wisteria growing on trellises, yellow climbing things for children in a playground, a red car with lots of rust and yellow flowers around it. • I'm outside Saint Catherine's Roman Catholic Church, on Middletown Road, after just being kicked out of a wedding practice. I tried to explain to them, tongue in cheek, using Latin. I was there selling our project, but think inadvertently I might've mentioned that my underwear was too tight, or something along those lines. Anyhow, they kicked me out, and I'm now heading back to the meeting place. • I am currently here in Lobotomy, New Jersey. Yes, Lobotomy, New Jersey. The sky is pink, the ground is blue, and the grass is purple. Yes, you heard me correctly, purple. There are several zombies walking around—one in a T-shirt, one in a bikini top, one with a little girl, and, uh, several with four noses. Yes, you heard me, four noses. It is crazy, but ya know, again, Lobotomy, New Jersey. Everyone forgets where they

are; they forget everything. • I'm in the marsh, down off the road below the Oceanic Bridge, beside the Navasink River. The tide is out, so the grass is exposed, along with the mussels and the shells and the rocks. I'm alongside the deck, behind the fancy restaurant whose name I won't mention. • I see a "Left Lane Must Turn Left" sign, a red light, four cars. I see a big brick wall, a push button for walk signal. • Hello. We're calling from the Ocean Club, which is in Atlantic Highlands, and it's dockside at the pier here. It's a really nice restaurant, raw bar, and cocktail lounge. It's starting to fill up with the Friday night happy-hour crowd. It has lots of outdoor seating, really nice awnings, and it looks like a nice way to start the . . . • I am here to tell you that optometrist John Colini, Elaine Goldberg, Susan Lennon, Ron Lennon, are all here in the Holmdel's Kohl's Plaza shopping center, ready to give you an eye exam. And when you're hungry . . . • I'm reporting from in front of the

Whole Foods Market, in the Holmdel Town Center Shopping Center, in Holmdel, and it's very quiet inside. Heard this place was mobbed a few weeks ago when it opened, but it's just like another market today. I will say it's very cool inside, and it's still about 105 degrees out here in the humidity outside. • Hello. I am at Lake Lefferts, in Matawan, New Jersey. I'm sorry; I am on Route 35, going south in Keyport. I stopped by the Shore Motel, and I found that pretty interesting. I know that a whole lot of homeless people are in there fixing it up. That's my son in the background playing the guitar . . . [Sound of humming]. • I am calling . . . (Vanya stop!) I am on 35 still, driving—traveling south. It really is a hot, beautiful, windy day. There's a lot of action going on at five o'clock in the afternoon. Well, have a good day. • I'm in the town, and I see Dirt Bag Dan, the wrestler— and we just videotaped his car. And it's, uh, he has bumper stickers that say, "SendaDirtBag.com," "Don't Get Even, Get

Ahead," "[unintelligible] Eat Me," "No Fat Chicks." Fat chicks have two stickers. Human League sticker . . . • Hi! Well, I'd like to just tell you that this most amazing dance occurred in front of the 7-11 between the Doughnut and the Iced Tea; and the Iced Tea asked the Doughnut out, but he didn't really want to go. • Hi, I'm just east of Little Silver train station. I seem to be in a neighborhood where I am not welcome. There seems to be nothing but "Private" signs and "Keep Out." I don't feel too welcomed here; I am going to keep on going by. It seems to be a very wealthy neighborhood, which is why I don't belong here. So, I am going to get out . . . now. • I'm on Route 34, in Colts Neck, at Delicious Orchards, at the meat counter, and looking at all the sausages, and the place is busy. I'm waiting in line; my number is three. They're on number 99 now. It's wooden walls around, and plaster walls, and glass cases with food in it, and fruit piled up, and vegetables, and tile floor, and

  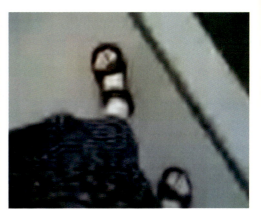

refrigerator cases, and carts rolling around. And [unintelligible] should be here . . . busy outside on the road, and that's it. Looking at the chopped meat, and there's the butcher. Three! [Background sound] I don't know why this is going on and on. Oh! My number came up. (Yes, I would like a quarter pound of this Westphalian smoked ham, please.) So I'm just getting my order here at the glass case, looking at the meats in the glass case here, standing next to a woman wearing black. This should've been more than 20 seconds here [background sounds]. So, it's 5:46; it seems to be going on and on here [background sounds]. . • O.K., I'm on Highway 34 in Colts Neck, at Delicious Orchards, and I am standing at the canned soup area with the corn chowder and the gazpacho, New England clam chowder and French onion soup, lentil sausage soup, and bean soup, and not too far away from the rice selection and the hot peppers. I am walking toward the checkout stand, past the

pasta and past the blueberries. The people are rushing around with their carts in this building, and I am about to check out. Oh, the flowers; flowers are here, too. People are getting ahead of me with their carts, and the sunflowers here I pass by . . . • I'm standing at the checkout stand, still—people in front of me by the dairy section, between the dairy and the flowers. Register 13 is closed, and I think I'll go to register 10, and that's opposite the cactus, which is in front of the eggs; and maybe I'll go to register nine—less people there. The man in front of me is buying celery, and the lady here is buying apple cider. The conveyer belt is about ready to take my stuff. The guy ahead of me bought $40.90 worth of groceries; my stuff is going to be considerably less. He's got more than celery here. Bananas—lots of bananas—and green, hard green bananas, and cantaloupes. The bagger is putting the cantaloupe in the bag now, and that's about it. The [unintelligible] look good; so do the

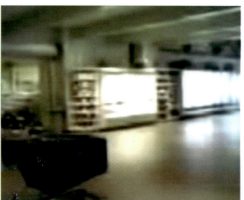

cream puffs. There's a bunch of candy here. I could have some, but I don't think I will. Just waiting my turn at register nine; it's a slow one. I should've checked out at 11. • O.K., Delicious Orchards, Highway 34. I'm sitting outside looking at the white sign with the black letters, and they have five hanging plants underneath it—impatiens, I think. The parking lot is emptying out because it's close to closing time. And little birds are hopping around on the parking lot to try to find any crumbs anybody left from their coffee and donuts, 'cause they have an outside area where you can buy pastries, and coffee, and donuts. And it's a hazy, hot summer day, and I am underneath a fruit tree before I am about to get on 34 to get on 537 to come on back. This is a lot longer than 20 seconds, and I don't know how long I'm supposed to talk. I bought some ham, and I'm sitting on a park bench near the parking lot. Seems like this goes on forever—so that's it; that's what it's all about.

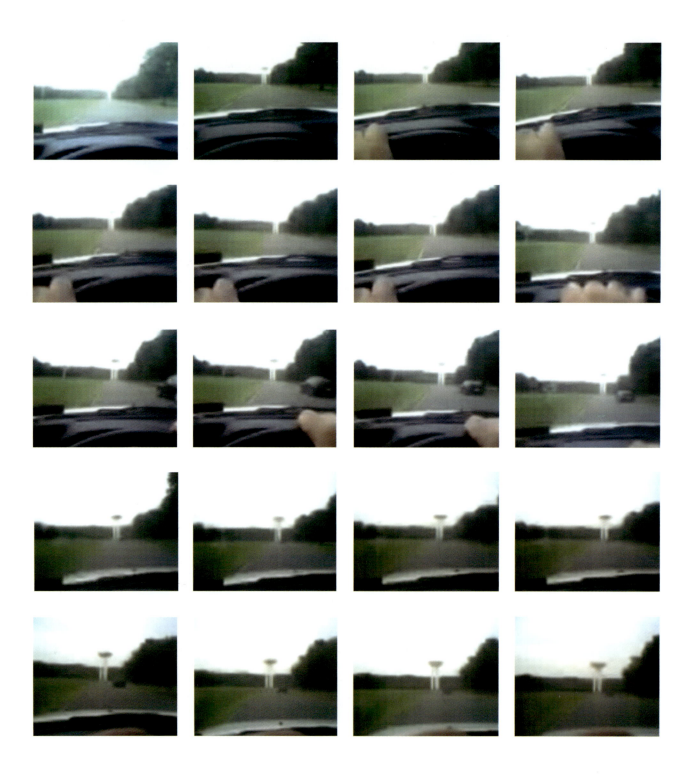

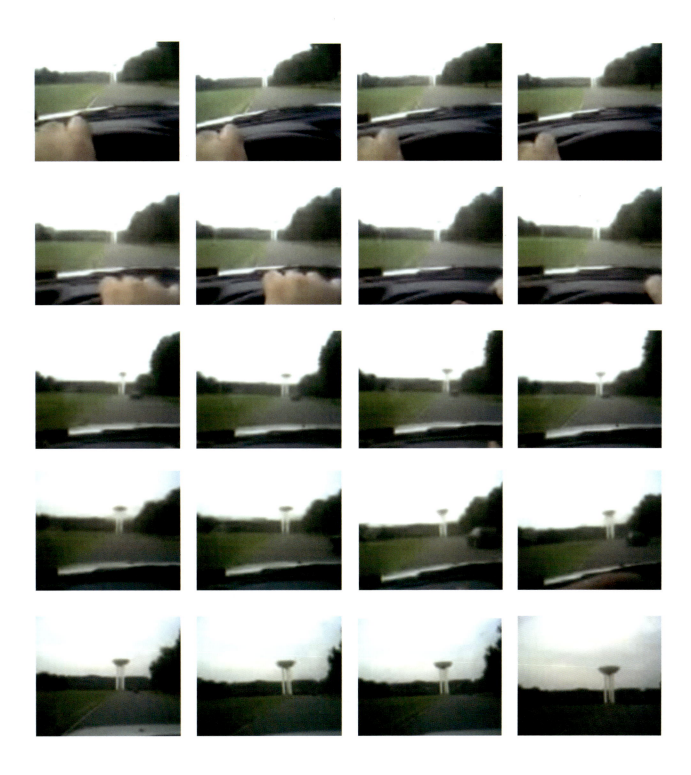

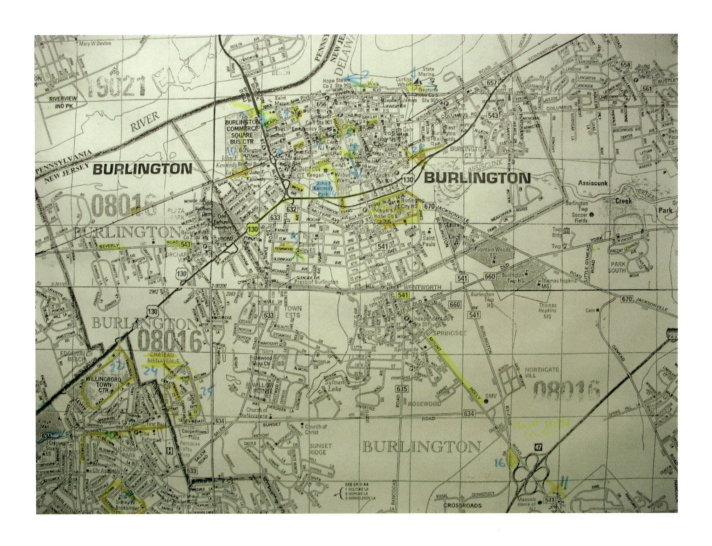

Metal giants stand above me, looking out, carrying power. Look at the elms. The energy, the source: It makes our country run strong. • I'm standing on the corner of Brown Street, just at the base of the Burlington-Bristol Bridge. I see a beautiful water tower, and a very well-kept garden with lots of American flags, and a tollbooth. It's very busy at the base of the bridge. On the other side, near the water, there's hardly anyone there—no cars, a little bit of a factory that I can see. • So, we are headed to the cemetery—not too exciting. • Hello. I'm standing in front of the [unintelligible], and children are flying a kite. And we're at the end of High Street, where there are some flowers and a painted bird, people walking around, and that's about . . . and the wind is blowing and that's about it. • I am standing on the corner of West Broad Street and Wood Street, looking down into downtown Burlington. You can see the top of the tower of an old church, the train station; very historic town. Looking on

Wood Street is historic London neighborhood—narrow streets, brick sidewalks, an old cemetery with an old iron fence encompassing it. Looking in the west direction looks like basic town, nothing of any consequence. Looking in the . . . • I am standing under the Burlington-Bristol Bridge. It's very quiet and peaceful here. There's a big log fallen in the water, a big tree. An alarm just went off on the bridge and broke the silence. Other than that, just a few boats calmly going down the river. No cars. • Well, the escalator is not working, so you run to the other end of hallway, and there's no elevator there. Look! There's a stairway that leads down into the fountain. So, you walk around the fountain—you're going to be here all night—and you come all the way back down because it was such a shortcut to take the escalator. This is a mall of many adventures. You walk around, and who do you see? Momma-and-poppa stores of all shapes and sizes. • I'm standing on Mitchell Avenue,

where there are lots of signs: "Birds of Paradise, 1,000 Exotic Birds." And you can buy a fish sandwich—fish with fries, ribs with fries—or you can buy the whole stand: "For Sale." You can call Eddie at 386-3516. I look to the left: There's the Love Shack—adult novelties, triple . . . • Hey, we love our customers; our customers are A-Number-1. I mean, like, they have made us the number one jewelry store in this mall. —Tell the story behind this ring here. —This ring was purchased actually from a friend of mine who . . . They melt gold down and separate the diamonds from the gold. So, he bought it, said he had a nice ring. So, I bought it. Good price? Uh-huh. Five hundred dollars. O.K. —Can I try it on? —Sure. • Hi. I'm at J.F.K. Park, in Burlington, New Jersey. The park is not much to see. John might be a little disappointed to be here. There are two little boys or two kids swinging on the swings, quarreling about something, but it's very calm. A few birds, and the park hours sign

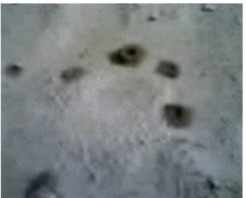

just standing there. • I'm standing on the corner of High Street and East Broad, and I am observing High Street—lot of traffic on High Street. I'm standing across from the River Line Café, named after the River Line train that goes through here every seven minutes. I'm standing in front of a display of flowers and American flags. This is quite the small American town. I'm standing in front of . . . • I'm standing on the corner of West Broad Street and Talbot Street, across the street from Saint Mary's parish office. There's a man walking his dog through the cemetery. There's another man sitting on a park bench reading a newspaper over the garbage can. People are walking along the street, coming from the train, walking home, I presume. • Hello. I'm on the corner of Mitchell and Williams. I see cages: cages for dogs, cages, cages, cages, empty cages, cages, cages, cages, cages, cages, yellow garbage can, yellow garbage can, more cages, green cages, black cages, silver cages,

white cages, and glass tile windows. • I'm standing out, looking over the river. The water's moving because the wind is blowing. And there's a big boat coming down. I'm not sure, it must be something. And there's a little bit of sunlight on the water. Now, it's getting nice and breezy, and the kids are still flying their kite. • I'm at State Marina, Curtin's Wharf. This is a nice-sized little town marina. The families would come and, like, take a boat ride and eat and drink here. It doesn't seem very active right now but [unintelligible]. I saw people try taking boats with kids and eating here. • I'm in Beverly, on Riverbank Avenue. I'm looking at the Delaware River. It's very gray here, little bit smoggy. There are, uh, some racing boats and there's a boat in front of us that says "Fast Tracker," but the engine's off. They're just drifting. You can hear the waves. It's pretty bucolic, clouds, a little bit polluted. • Uh, being at the Curtin's Estate Marina, Marina Curtin's Wharf for a while, ah, I could

  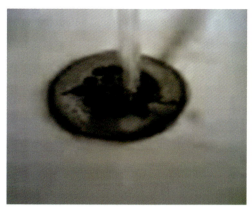

start imagining that this place had more color and more live atmosphere in the past, and when you're up crossing the little bridge . . . • Hello. I'm on the corner of High Street and Federal Street, in Burlington, and it's an area with lots of lovely, ah, mixture of houses and traffic and overhead cables. But it's very busy and looks very interesting to me. • I'm standing in a park across from Pearl Boulevard, looking across the river to Pennsylvania, and a sailboat is going by. • You can hear the Delaware River lapping on the shore here. I'm on Riverbank Avenue, in Beverly, New Jersey, and it's kind of nice. Although to the back of me—I'm facing west—to the back of me, it's kind of some derelict buildings. I see some seagulls flying, some terns. • Sitting in the playground across the street from the school, on the corner of Ellis and Pearl. Sitting on a rocking horse and looking down at the ground. There is not even dirt on the ground in the playground; it is all chomped up bits of tire.

Funny to think that even dirt is . . . • I'm just at the base of the Burlington-Bristol Bridge, and I just saw a fabulous sight. I hope you get it on the video. The whole bridge, the alarm went off, the traffic was backed up; the bridge lifted up, and this enormous barge just cruised through with a big crane on the back. The bridge must have lifted up about eight feet, and the barge was enormous, tilting to one side with this big load on the back of it. So, the video should be coming in soon. • There are many, many, many ducks here, and some water lilies growing in the water. It's actually continuing along the river; and as I'm walking towards west, there's some housing starting to be showing up. • I'm on the corner of East High and West Broad Street, and High Street and Broad Street. I'm looking at a bird statue—a statue of an American eagle—and the feathers are made out of parts of old baseballs, and it's called, "Casey at the Bat." And it's in front of an Italian restaurant. • This is J.F.K.

Park, in Burlington, New Jersey. The two boys who were on the swing just said they tried to talk to them; however, the park is never really empty. Now, a lady with a dog came into the park. Lady's trying to get to catch the dog with a . . . • I'm in Beverly, New Jersey, at Riverbank Avenue. I'm facing a power pole that says 663PE in metal. And there is some, a poster for a Saturday night revolution, and in front of that there's Dunk's Ferry Bicentennial Memorial. It's one of the first ferries in the United States. There is a plaque here, actually a monument. • Walking down the street, on Pearl, the wind blowing through the trees. It looks like it's going to rain. No kids playing outside—really quiet. Think it's going to rain today; getting cold, weird wind, little bit of a warm breeze. Nobody out; nobody around. • I'm standing next to the Burlington County College and the plaza with an amphitheater. Right next door is a large . . . what looks like an old movie theater boarded up,

pretty much abandoned. • I'm in front of the Patio Knitting Company, on the corner of Williams and Mitchell. It's a beautiful, old abandoned building, broken glass windows. I'm also standing across from Bird Paradise, on Route 130, which doesn't look like such a paradise for birds. • I'm standing at the corner of High Street and West Pearl Street, in front of a sign on a red brick building, and there is a white archway with a beautiful black door. And I'm walking down a cobblestone sidewalk at the moment, and that is it. • I'm in Burlington, New Jersey, on Riverbank Avenue. I'm looking at a tern on an iron fence. There is a mallard duck in the Delaware River. You can hear the Delaware River lapping here on the rocks. Behind me are picnic benches and a park that no one's in, although there is trash all over the place. Kind of an empty desolate area here. There is a car waiting for me with the engine running, because it's hot and humid, and the air conditioner's on. • I'm in Beverly,

New Jersey, standing on one of the main streets. There's many American flags flying high in the sky. I'm outside the National Cemetery. It's such a beautiful scene here on a nice, hot summer day in New Jersey. • I'm on Pearl Boulevard right now, right down the street from York Street. There is a woman sitting on her stoop, talking on the telephone, an old rotary phone that she's dragged out from her house. She's barefoot; she's drinking soda out of a McDonald's paper cup; and she's smiling at everybody that walks by, waving "Hi." She looks like she probably lives in this house. • I'm in the Blockbuster Video action section in Liberty Square. We have "The Sixth Day," "Twelve Monkeys," "The 13th Warrior," "15 to Life," "24 Hours in London," "100 Kilos," 187. . . • Standing in a dirt lot between two three-story, brand-new housing developments, neither of which has been sided. The windows have been placed in, and there is construction material and debris, waste

  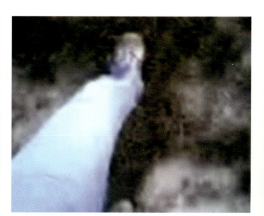

material all around. Surrounding that is a multitude of parking lots and a few trees here and there. • I'm looking at a chicken bone, on Williams Street, next to a piece of aluminum foil. Now, I'm looking at weeds as they go downstream; I'm leaning over the bridge. • I'm on the corner of West Broad Street and Ellis Street, facing west. We've left the downtown Burlington area. Directly in front of me is the Burlington Water Tower and the access area for Route 130. Across the street from me are houses, brick houses. Someone's asking me what I am doing out here, through the window. In the other direction, towards the east, towards Burlington, people are on the sidewalk, the brick sidewalk. And toward the north is a police car with his lights on. • On location: Mitchell and Williams, partially sunny, high humidity, chance of rain. We're in front of the old recycling plant; blue garbage cans as far as the eye can see. "For Recycling Purposes Only." Come on down, bring down your

 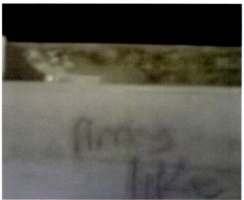 

recycling: glass, plastic, metal. • I'm on the corner of East Broad and High, and I'm watching people line up to board a train and walking up the ramp onto the platform. I see a lot of men with green jackets—looks as though they were working in construction of some sort, with bright neon-green jackets—and they are all walking together, standing on the platform waiting for a train. • I'm on High Street, Burlington, looking at a wonderful old building with magnificent old bricks and a shingled roof. It must be the oldest building in Burlington, and it's really lovely. I love doorways, too. There are lots of good doorways. That's all. I don't think I have anything much else to report. Gosh, I can go on. Well, it's very interesting, but I'm going to the cemetery next that looks a bit more . . . • I see a policeman, who has just pulled up behind me, because I just pulled over to the side of the road. Bye. • This has got to be one of the most patriotic places in New Jersey. Here, on Pearl Street, I'm looking

around; I think I've seen more flags in the last few minutes, walking down this street, than in my entire life combined. Standing at a Veterans of Foreign Wars place now, looking at even more flags. I didn't think that was possible. There's even a banner of metallic flags hanging in front of the bannister leading up to the front . . . • This is J.F.K. Park, in Burlington, New Jersey. The park is next to a highway, where the most frequent visitors are the cars. But, right now, there is also a family—the dad mowing the lawn, the mom picking up the cut grass with a broom, and the kid following the mom picking up whatever . . . • I'm standing on Mitchell Avenue on the bridge, the Burlington County Bridge, that was reconstructed in 1979 and seems to be the gateway to the neighborhood of New Yorkshire, which was established in 1696. So, I guess that until that time, it had been cut off from the other side of the river. I see some boys playing basketball, and I see my car, which is good. • I'm

on High and East Broad Street. I'm watching a woman in an orange T-shirt and tan slacks sweeping the platform of the train station. Beyond, a long Caprice wooden station wagon with a loudspeaker in the back of the car, with the back door open and the back window lifted, and the train approaching the platform. And you can hear it in the background. This is a long train. • Standing at the southwest corner of a strip mall outside the new Happy Family Buffet. Seems to be open though no one's eating. Across the street—another dirt lot ready for construction, foundations poured. Beyond that, a highway, a Burger King. Around me, more and more parking lots. • So, I'm standing in front of Beth B'nai Israel Temple on the corner; well, actually it's on High Street. And there's beautiful stained-glass windows with stars in them. And I'm still walking down the sidewalk, and there's flags in the windows probably left over from the Fourth of July. That's all I'm going to say. • I'm standing in front

of a trailer. "For sale, call 636-3516," if you want to buy it. • I'm on Riverbank and Wood Street, walking down the boardwalk. There are a couple of boats coming very fast towards me, speedboats. But there's not a soul along the whole boardwalk for nearly two miles. Now, I am just seeing two people walking out, and they are leaning over the balcony, talking, chatting by the . . . • We're in Toys-R-Us, and the exit signs point to the wrong doors. In case of fire, we're going to die. The Dino ride is out of order. • I'm standing on Mitchell Avenue, outside one thousand . . . Thousands of Exotic Birds of Paradise. It looks like a package exploded: There are packing peanuts all over the parking lot and outside—those white foamy ones that look kinda more like a cheese doodle than peanuts. • Hello. I'm reporting from the Acme in Liberty Square. Checkout line number one is closed. Checkout line number two, open; number three, closed; four, closed; five, open; six, open; seven is closed; eight,

closed; nine—handicapped access—is open; 10 . . . • I'm at 605, Route 135 South. We're at the Love Shack: adult novelties, DVDs and videos, private viewing booths. This is great; seems like there's plenty of parking here, moderate traffic . . . • We're at the grocery store in Liberty Mall, and I feel free to buy bundt cakes, bundt cream cakes, for $3.49. • I'm on the corner of Union and York Streets, standing in front of the Church of Holy Light, Church of Jesus Christ, and Holy Light Christian Academy. There are some fire engines behind me, so you may not be able to hear me too well. Probably can hear the sirens right now; they seem to be coming down York Street. There's a man walking down the block wearing a long, white T-shirt hanging around his baggy jeans. He waved to the firemen. It's very loud. [Sound of sirens] VERY LOUD! • I'm now standing at the corner of Plaza Circle and Millennium Way, looking at the abandoned movie theater, which will soon be 90,000 square

feet of class A office space. Across is another parking lot and some dirt piles. I can see a diner, the Golden Dawn Diner. • Ian and I are sitting outside. It's a really hot day, and the concrete is really fucking hot, and we're trying to work on Ian's car, Alan's car. Ian, what do you know about cars?—Absolutely nothing. [Sound of horn honking] I know computers, and I work on computers all day.—Do you get to go see the ocean at all?—In summer? A couple of times on the seaside, that's about it. • I see a sign that says, "Buckle up—someone loves you." I'm in some sort of industrial area: Craft Fair, some other companies, Kenneth Cole; strange. • I'm standing on the corner of West Pearl and the river, and an incredible amount of fire engines went by. We saw one from Willingboro that went, turned around—the whole fire department—then went back the other way. I think he was lost. We're now coming up on the fire, and it's about four fire trucks, all flashing lights. We're passing the church and there seems to be a real emergency fire in the old . . . the brick building, that great brick building. • Steinman's Frame

and Wheel Alignment, a line of blue plastic recycling garbage cans, black plastic bags caught in a tree in front of the old knitting factory, little shards of glass left in the knitting factory. A guy is riding his bicycle by. • Standing next to a yellow gasoline-powered air compressor and a pile of 2 x 12s, as three guys toil in the heat to frame out the first floor of what looks like another housing, multistoried housing development. It's hot. • Well, I'm in the middle of a big drama, beside the river-side on the corner of . . . • There is a big fire on the corner of Pearl Street and . . . • We're here on the corner of Bucknell and Brooklawn Drive. It's pretty quiet on this block today; it's about 90–95 degrees. There's a house here that looks like they recently had some tree trimming done; and other than that, the block is pretty quiet. • O.K., as you can see we're on the corner of Somerset Drive, right on the side of Willingboro Christian Assembly Church. It's about 90–95 degrees today. As you can see, they are doing some construction in front of the church, and it looks like all is calm at this present time.

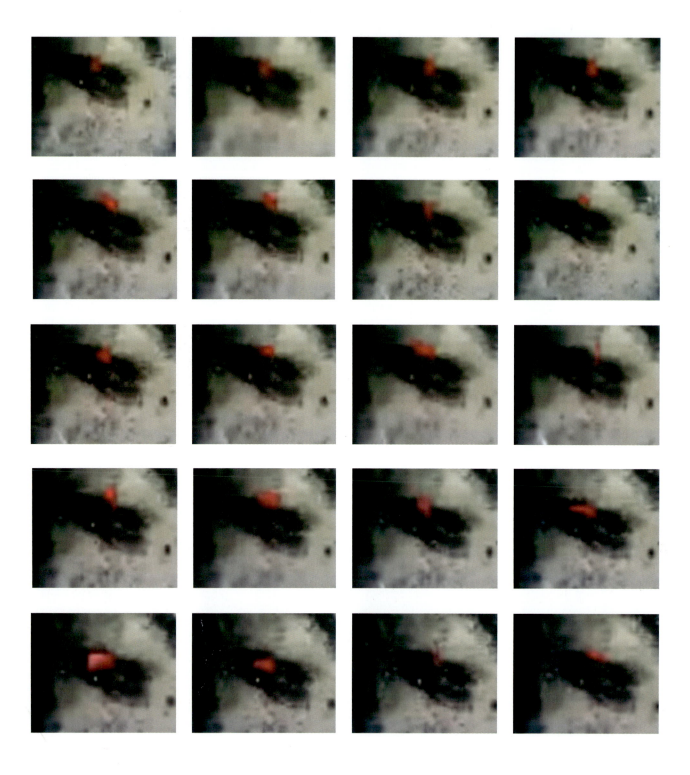

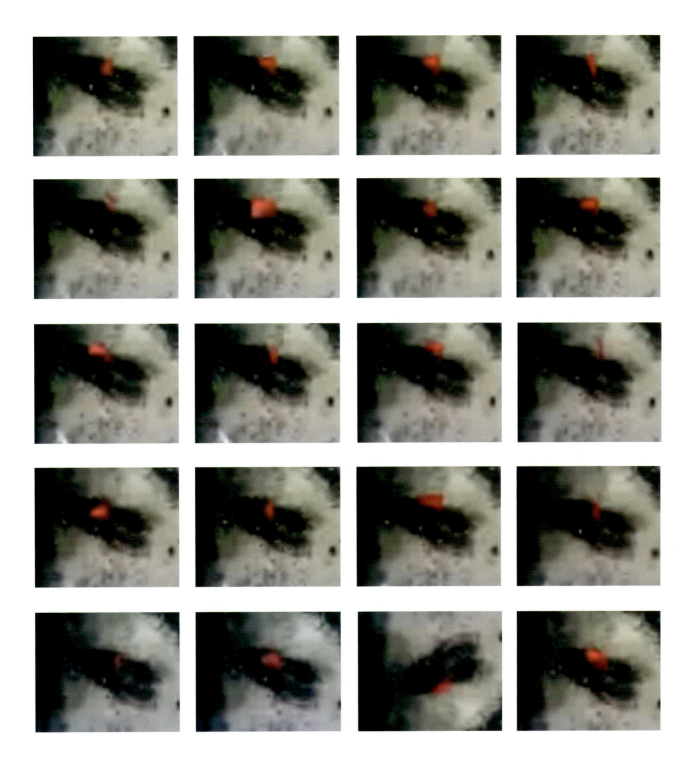

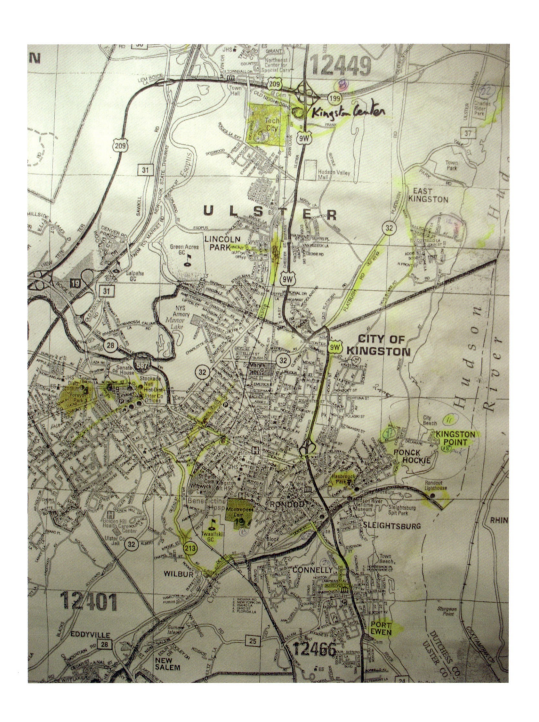

Hello, this is Pauline. I'm standing across from T & T Surplus. This is the Canal Street of Kingston, where you can find every kind of surplus that you might like. If you're an installation artist or a mechanic or a technician, this is the place where it's at. Down here on the Rondout. • Now at the Kings Inn Motel, with hourly rates, if you want to take a quick nap, or do something else that takes less than an hour. Kind of a run-down part of town. Old cash registers . . . • The cemetery is so quiet, except for the gentle buzzing of a plane up above and the sound of someone chopping down a tree. My dog, she's running lightly through the tombstones chasing after squirrels. So much peace, quiet, rest. • This is Cynthia reporting from Devil's Lake Road and John Street, in this town of Kingston, New York. I see a Costco van in front of my eyes, and to the right are many chicory flowers, and down the street there is a "Children at Play" sign, and across the road

is a small, quaint house with green shutters. • I can see a tugboat spraying multiple shoots into the air of water, passing by a lighthouse. I see and hear the water in front of me, and a piece of driftwood is toiling in the water. • I am standing at the Dietz Stadium, in Kingston, New York, and there is a very large track here with a red, rust-colored running portion, and then a thick green center for field hockey, and football, and all sorts of other sports. There are a lot of people who are getting out of their cars and coming over to run the track. Someone is coming in a yellow shirt that says "Ray's." And there's advertisements around the outside of the local food establishments: Rusty's, Harry's Deli. There's one, two, three American flags. • Hi. I am standing outside of T & T Surplus down in the Rondout. I love to go here because there are so many wonderful things. All kinds of art, as well as surplus electronics and machine parts, and technology, and all sorts of

things. • Hi. I am currently at a crossroads in Hurley. There is not too much going on here. Lots of American flags, and lawn-gardening trucks going by, corn and old stone buildings. It's a lazy town. • Sun Street and Silvertone Lane, in Kingston, New York. I see an American flag. I see the Miller's home and a "Beware of the Dog" sign. I see a dead end on Silvertone Lane. I see another American flag. I see many geraniums. I see a number of homes on the . . . • Kind of a gray overcast moment in consciousness of the human species. Trash on the sidewalk; happy hour at Broadway Joe's. And skittering down the pavement, cigarettes flicked out the window, sunken down sideways. Kind of makes one wonder what's going on. • Oh! I'm standing on the pitcher's mound, looking out over the playground; and I'm throwing the pitch, and here's the windup. Actually, this is all a lie; I'm just looking at the tennis courts and woodchip pile, not even pitching at

all. I'm just standing here. • I'm standing by the railroad tracks in Saugerties. And there's an elephant here. It is a big circus prop. An old merry-go-round next to it. And there's people unloading furniture into a building, and there's a herd of goats. And this place was supposed to be chosen for a giant paper-processing plant—newspaper-processing plant—that got kicked out of Saugerties, and now the goats all have cancer. • I'm standing in Saugerties, on Route 9W, between Glenerie and Barclay Heights, in the parking lot of Barn Raisers. I'm looking at some outdoor furniture, some of them with horses and wolves and bears, wishing wells, gazebos, sheds, garages. They also have a gift shop. Next to it with some construction under way for a senior citizens' condo complex. Traffic is kind of heavy. It's an overcast day, a little bit windy. • I see a lighthouse with flags waving. I hear a tugboat whistle, I smell the fresh water, hear the leaves whistling in the

wind. I'm walking on gravel. I hear crickets. I see another duck. • Reporting from Kingston Point Park, walking around and struck by how short the grass is clipped and how, when I look around me, I see all the colors that I see are all green, except for the waste receptacles, which are a blue one, a yellow one. Kind of artfully arranged, actually, kind of strange, with a long, yellow foam padding over the top of the fence around the baseball diamond. Sort of adds a nice little accent, almost. Someone incongruously placed a dumpster just out in the middle of the field. • I'm at Kingston Point, off of 9W. I see a truck driving by with three kayaks; it's a Dodge pickup. I'm seeing a park. There's a bathroom station next to a couple of really industrial water tanks and some sort of abandoned stockyard. Then, panning to the left, there's . . . • Hi. I'm at Belknap Lane, on 9W, in Saugerties. There is a house for sale by owner: "Must Sell, Must Sell." Down that way,

  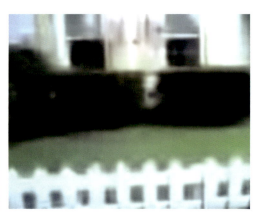

9W north on the right hand, east side of the road, the house for sale by owner, "Must Sell, Must Sell." • I am currently off of Green Street, in Kingston, in a parking lot. There is an Emergency One truck, with emergency mobile vehicle on a "Permit Parking Only" next to a Volkswagen Passat, which is very patriotic with flag paraphernalia. And there are other people in the parking lot sitting in their . . . • I'm at the railroad yards, oxygen acetylene. I'm peeing on a big propane tank. • The Twaalfskill Country Club golf course. Looking at cloudy skies, a white rail [unintelligible], a red flag, and the timeless view. • You're not going to believe this, but no wonder there is no one in Hurley, because there is a wedding at the Hurley Reformed Church. There is a lovely lady in a beautiful fluffily ruffily white dress, lots of candles going, and organs playing. Pretty exciting day. • I am at the Forsyth Park Petting Zoo. And I am looking at a zebra finch, several zebra finches, and

some very sad tropical birds, one of which has its feathers plucked out all around its neck. And there is one child petting the goats. • N1693R, red, blue. We're at the Kingston Ulster Airport. Green trees; a brunette is walking slowly down the driveway. There are yellow [unintelligible] signs preventing us from going over to the airport. There are approximately 20 planes here. • I'm on 9W in Barclay Heights, with an old house which has all of its Christmas lights still up. It's yellow, with blue windows, next to a very tall willow tree. • (We're doing video, too. O.K. I'll talk about it, but I was wondering about those kids, too. Oh! I'm on, and I don't even know it.) Well, we're just sitting here in town with this nice lady at 123, but here talking to [unintelligible] Radio Shack. The local kids already took off in their little cars, and they won't talk to us either, but they probably had a party to go to or something. • This is Pauline. I'm standing outside on Abeel Street facing

east, looking at the old Forst Meat Packing Plant that's slated to become the Noah Hotel, a sixty-room facility that will change the face of the Rondout. I hate to see the old packing plant come down. In a way, it's got a lot of style and character, even though it's on its way out. • I'm Dave, and I have a maintenance business out here at Kingston Ulster Airport. A maintenance business works on general aviation airplanes that are based here at Kingston Ulster, and other planes that come in from different areas of the Northeast. We work, basically, on piston single- and multiengine airplanes. We do inspections and maintenance and repairs as needed. Thank you. • Overlooking the parking lot at Kingston Point Park. The parking lot is almost completely flooded. There is a huge puddle with a group of cars. Local high school kids seem to be having a great time going at top speeds through this puddle. And then they go around the parking lot and do it again,

 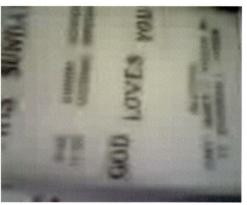 

and again, and again. Looks like they are having a pretty good time. • I see railroad ties. I see sea chestnuts, spiky and sharp. I see driftwood. I see sycamore and locust trees. I hear waves hitting the rocks. I see water [unintelligible] floating on top of the water. • I'm walking at Kingston Point. I see a man with a white shirt and bright green slacks wearing sunglasses and waving a cane. It looks like he's actually blind and is walking along the highway—a bit dangerous. And then there's also several cars passing me by right now—an enormous sort of pond created by the rain. The man is still walking along and narrowly being passed by cars. • I'm in front of a home for sale that is listed by Paula J. Kitchen, a licensed real estate broker at 803 Ulster Avenue, Kingston, New York. Her number is 339-9800. Once again, this home is for sale. • Mapweed, Queen Anne's lace, butter-and-eggs, maple tree, all trying to grow by the railroad tracks. A little

  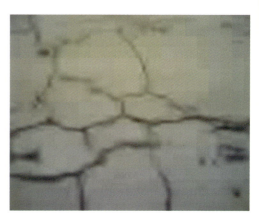

cedar tree, bridal wort, Aaron's rod, meadowsweet. A train just went by a few minutes ago; graffiti and screeching wheels, just a few cars on it. • (I'm next when the music stops. I'll pull over here.) There's a barking dog [sound of barking]. You hear the dog? I am at Lucas and Lafayette Avenue, and there's a big white dog chained out under this tree. • Why are we wandering these parking lots? Seeing the shuttle buses come and go, taking people working at the hospital. Hard to fathom railroad tracks running here and there. • They are actually practicing, and they are not very good, but, uh, we have high hopes for them here. There they go again [sounds of whistles and voices]. They're running. Not doing a very good job of anything. They're losers. They were about two and ten last year. • Still at the Forsyth Park Petting Zoo. The peacocks are strutting. There are birds inside a cage and birds outside the cage. There are very few people here. Two

trucks parked by the baseball field. And there are machines for putting a quarter in to get feed for the goat. • I'm on Route 9W north. I'm looking west at the Catskill Mountains. The mountains are dark purple blue, and there are many clouds rolling over them. It rained here earlier. There may be more rain. The clouds are moving from left to right, from south to north, toward the northeast. It's cool and windy. • Lots of cars going down Broadway. A man coming out of a deli, getting into a red Chevrolet, getting on Broadway. • Hi. This is Pauline, and I'm watching the boats. People getting on and off the boats, on the Rondout, looking down towards the Hudson River. There's a big boat about to take off on a cruise. There's a [unintelligible] in the harbor, in the park, by the artist Susan Togut. There's a lot of action down here and all sorts of new developments. • Ah, the luxurious fragrance of hassle-free roses. I'm standing and walking on cobblestones: in loving

memory of Gail Beck, in loving memory of Louise Kelly, in loving memory of Mr. & Mrs. Michael Kennedy. • Hi. We are at Green Avenue, or Street, and the Holiday Inn. And we're passing Super 8, and there is more malls. There is also an art framing supply; there's a lot of art-craft stores around here. We're also now at the diner, and there is a nature walk coming off the back of the Holiday Inn. • I'm standing on Route 9W next to a "Saugerties Welcomes You" sign. We're in a field by the cemetery. • I'm looking at little purple flowers, and I'm standing next to a yellow fire hydrant. And on the ground is a cutting board, one of those that has a hole in it you can hang in the kitchen. It's just lying there next to the yellow fire hydrant. • I see loosestrife, goldenrod, and morning glory flowers. I hear tugboats whistling behind me. I see the bridge ahead of me on the horizon. I hear the breeze in the trees. I see the old wooden piers poking out of the water of the

Hudson River. • [Sound of background conversation and a baseball game] • Standing at the corner of Delaware and Lindsley Avenue, in front of the Fifth Ward Honor Roll, World War II Veterans Memorial. The memorial is flanked by two bullet-shaped shrubs. That has a plastic floral wreath—red, white, and blue—attached to the monument with a suction cup. • I'm still on Abeel Street with Pauline. I'm looking and seeing a couple of steeples—the seven church steeples on the hill. I'm at the foot of the bridge, and I see Burton's Car Repair, and I see the West Strand Mall. And the traffic is quite heavy at this time. • I'm on Leggs Mill Road, looking at the local DMV, which is full of cars. Across the street there is a lovely pastoral scene and lots of cars going by. • We are at the O & W Rail Trail—the D & H Canal Heritage Corridor—and it's a trail that goes along the tracks, with power lines above us as well. But it's a little nature trail and we're sending video.

It's a lovely trail. • I'm in Forsyth Park. I'm standing on a concrete platform that doesn't go anywhere. There are three teenagers here, a pampered child throwing a stuffed animal. Another small baby is kicking in its stroller because she sees a wooden playground structure. • I'm standing here by an abandoned railcar, rusty with vines growing all over it. Just confronted by Kingston Hospital security man. Seemed threatened by my presence, said, "Ma'am, you can't be here. This is private property. You can't be taking pictures here." He seemed genuinely upset. • It's the Harley Road King. I bought a ten-dollar ticket, and I won a Harley. Thank you very much, Jean. That was Peggy with me, and the other young lady in front of me is Jean, and she's wearing a lovely necklace. • I'm at a place called [unintelligible]. It's a place where they make cross-expanded shale—uh, cement—for the roofs of sport stadiums, and the World Trade Center, which fell down,

and other places. And they're going to be trying to build the new World Trade Center, or whatever they're . . . Freedom Tower . . . out of some of this stuff. It's a very dirty, dangerous place. • Two men are walking out of the building. One has a lunch box and a rolled-up newspaper. The one man on the right is going faster than the one on the left. He leaves him behind and picks up his cell phone. He's using his cell phone now. He goes over to his car, puts his cell phone in his pocket, opens his trunk, gets into his gray car, says parting words to the man he left with. • There's an old, derelict, railroad service yard down here, with old engines just rusting with vines growing over it. Their horns now silent that were once blaring. Wisteria creeping in and out of huge diesel engines. • "Frida's, the Best Mexican Cuisine, A Taste of Mexico," "WKNY, the Kingston Station," "Broadway Lights Diner," "One Way," "Open for Business," "Regional News

Network." • Well, I'm on my way back, and I just passed the state trooper shop. They happen to be the F Troop, which I thought was kind of funny. And I'm coming across the Sky Top Motel, which is above a big cliff. Looks pretty old, like everything else around here. • I'm standing at the end of Corporate Drive, in the Kingston Business Park. The end of the road: It just stops. The double yellow line, the end of the asphalt, just turns into woods. There is no actual business that I can see in this corporate park, which is a little bit perplexing. There are many signs, a lot of orange posted "Private Property" signs, so maybe I shouldn't necessarily be here. There's a broken down lamppost; it looks like it's just been torn off of its mooring. • (When the music stops, speak? Not really . . . O.K. Esopus, O.K. I can't hear. Just listen . . . Talk.) We're at the Esopus Creek Bridge, Hill Street and East Bridge Street, in Saugerties. It's the watershed of the Hudson

River estuary, and the traffic is moving well across the bridge. • Hi. I'm at the corner of Hurley Avenue and Washington Avenue, and there's a very old fire station. It says, "Excelsior Engine Number 4," and it's yellow, and it's very historic, I think, or old. And it's got a person in front, and they're just smoking. • We are back in the car, at the corner of Washington and something at the Trailways Station, and we will soon be returning to headquarters, if we don't stop at the Stadium Diner and Restaurant, which is open. • We're standing on the bridge in Saugerties. It's painted light green. If you look across the road, there's very still water. There are a lot of lily pads on the lakelike area. On the other side of the bridge is a dam, and the water just disappears over the dam. In the foreground, you can see rocks at the . . . • I'm at the Trailways bus station at Grandma Brown Lane, and the Kingston city-bound E bus has just pulled up. But, mostly, it's like

I said: Trailways, and a lot of people with luggage, talking on their phones and waiting for their buses. A taxi just pulled in. There's a woman in a green shirt with a . . . • I'm on the corner of North Front and Washington, across from the bus station as well, but reporting from Deising's Bakery and Coffee Shop, which is open seven days a week. And it's got a windmill up front. And there's a man sitting outside eating doughnuts, and he looks like he is enjoying his doughnuts very much. • I'm looking at a red windsock with some wind blowing through it. I think it's supposed to help the planes know what direction the wind is blowing. And there is a circle of red and white rooftops—little tiny pavilions. I don't know what the circle means. I'm going to have to ask someone. And behind it is a . . . • Oh, so I'm standing here under statues of Elvis and Marilyn Monroe, on the corner of Broadway Joe's, next to the pawnshop and Kennedy Fried Chicken, across

 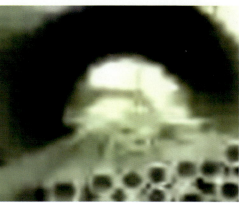 

from the Midtown Dollar Store. Children smoking cigarettes. Junkers driving by. • Star of David, a wall of hexagons, red bricks, blue cars, 231 apartments. • I am at West Brook and John Street, and we're going to the Historic Kingston Trail. And in front of us is very old homes, sort of Victorian-style. • I'm looking at a green building, and across the street is a brick building in great disrepair. On the second floor, there's a door leading to nowhere, and the door itself is boarded up by a board. There are green shutters on the building, and the porch is only I-beams. • Main Street in Saugerties on a Friday evening, and the things are heating up here. A little bit of smoking clusters outside every bar, and people walking around. And there is a cowboy, a bum, and there's a couple of kids throwing a balsa-wood airplane. It's electric, positively electric, here in Saugerties. • I'm at Main Street and Clinton. I'm going into the Main Street Market, and

they have various ice cream, deli meats, chips, cookies, sodas, household items, tea—also various assortment of Indian spices. Interesting . . . Hello, this is Pauline. I'm standing across from T & T Surplus. This is the Canal Street of Kingston, where you can find every kind of surplus that you might like. If you're an installation artist or a mechanic or a technician, this is the place where it's at. Down here on the Rondout. • Now at the Kings Inn Motel, with hourly rates, if you want to take a quick nap, or do something else that takes less than an hour. Kind of a run-down part of town. Old cash registers . . . • The cemetery is so quiet, except for the gentle buzzing of a plane up above and the sound of someone chopping down a tree. My dog, she's running lightly through the tombstones chasing after squirrels. So much peace, quiet, rest. • This is Cynthia reporting from Devil's Lake Road and John Street, in this town of Kingston, New York. I see a Costco van

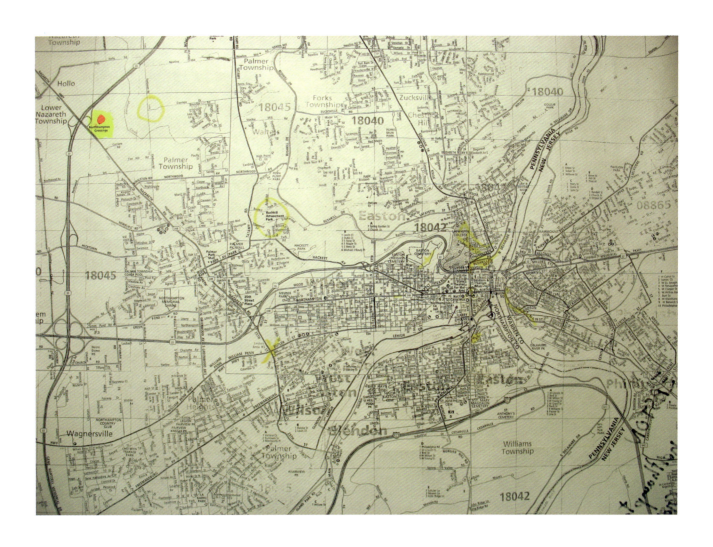

Hi. I was on my way to Klein farms, but I've stopped at a wrong turn, and I'm looking at . . . • So, there seems to be a number of people boarding horses here. And most of the horses are coming in from the rain, and they are going into the barn. It looks like they are getting brushed down . . . a little something to eat. There are a number of young girls here who are grooming their horses, and, let's see, about three more horses just actually came out to this big outdoor ring. • Well, now I'm at Klein farms, which is somewhere in Lower Mount Bethel Township, and I'm surrounded by big . . . • This is [unintelligible]. We're at the corner of Third Street and the Circle, and right now there's lots of people. • We're in Centre Square; there is classical music playing. I believe it's Beethoven. And there seems to be an awful lot of drunken, homeless people sitting around taking in the culture. • We're at Bushkill Park, where a woman has just asked . . . • We're at Bushkill Park, and a woman

is really upset that we're here; she may . . . • We're at Bushkill Park, where a woman has just taken down our license plate number. • The woman has just gotten back into her car, and she's putting on her seatbelt. We're waiting to see if she's going to come over to our car again. I'm looking at the Bushkill Park sign that we previously sent in video, and a rather large sign underneath that says, "Park Closed." Over and out. • Watching a man on a bicycle, with an orange shirt around his waist, passing in front of a white house with an American and Mexican flag hanging out front, oh, and an old Chevy Bronco, black. • Looks like the woman is going to leave us alone. She's driven back to the entrance, and she's waiting to make a right turn onto Bushkill Park Road; not sure if she called the police or not. Think it might be a good idea if we leave. • I'm standing equidistant between two buildings over 600 feet long in Lehigh Valley Industrial Park, looking

due south at another building that is probably another 800 feet long that is perpendicular to me and roughly at eye level with the roof of that building that is about 900 feet away. About a mile in the distance, I see a small hill—green—with houses, very few, on it. There's one truck, white, which is parked in the loading dock of the blue building on my right. • True Libras are simply unforgettable. They have a flair for the extreme, a sense of the dramatic, a need to express themselves publicly, and a penchant for hair extensions and caftans. Libras let the world know what they want to do and who they want to be. They can belt out a song, no matter where they are, in front of an audience. • Hello. We're at . . . Hello. We're here in Easton Cemetery, my wife and I. We're the only people here, at least the only ones alive, and walking among the tombstones. There are incredible carvings: obelisks 40 feet tall, a great bronze guardian angel. • Greetings. I'm on Hecktown

Road, inside a model home, where a bunch of houses are now being built. The home is very nice—a little bit expensive, but very nice. Beautiful views of the surrounding area, which, I'm guessing, used to be cornfields. There's a lot of homes going up here. • I'm looking across a drainage ditch at two cyclone fences running parallel to each other. They're about 50 feet long, and they're sheathed in a black material that retards soil erosion. In the background are three smokestacks that are probably close to 200 feet high. One is much fatter than the other two. • I'm at a three-way stop on Van Buren Road, near someone's farmhouse. A yellow Ford pickup, older, is next to a bank barn. Next to it is a small shed with an American flag wrapped around itself. • Easton, Pennsylvania—Lafayette College campus quad. Four white stripes; the Beatles crossing; behind them, students looking for other posters. • Lafayette College, looking south on Third Street towards down-

town Easton, towards the Circle, towards U.S. Route 22; overlooking the Bushkill Creek. • I'm at Braden Airport, and I'm watching two men with burgundy shirts talk behind a counter, through a window looking at a bush. Now, I'm looking at a sign that says, "Attempt to Enhance Security Procedures in Effect." There's an American flag, and I'm looking at a plane with burgundy stripes and the letters 09163Z. • I'm standing in front of the forks of the Delaware where the Lehigh River dumps into the Delaware River, looking at a poster of two mules and a man. They used to provide the power; the mules used to provide the power to move up and down the Delaware [River], the Delaware Canal. • All right, I'm looking at a handbag hanging from a chair. It's kind of a rolling chair with four little feet on it that rolls, and there's a flag on the chair and a man who is slightly overweight, who is talking to himself, who's wearing a bright pink shirt. • I am currently at Braden

Airport, in Easton, where they offer plane rides, aircraft maintenance, and sales and service. You can learn to fly here. And I am watching cars go by. And the clouds up ahead; they are looking kind of dark. • I am standing in Lower Mount Bethel on a farm, and I'm looking at about six young calves that I hear are about 3 1/2 to 4 weeks old. They still have umbilical cords actually attached to them. They're black and white or brown and white, and they have brown noses or pink noses. • The woman who is suspicious of us is sitting in her van, still, at the exit, making a left turn, probably waiting to see what we are going to do. • I'm looking at a piece of concrete that was left as garbage in an industrial park. It's about eight feet tall; it's slightly angled 30 degrees; the edge facing me is polished smooth. Against it has been poured a mound of tar. The mound of tar is about four feet tall; the wedge of concrete is stuck in it. • We're in front of the Pleasure and Pain, looking

across the street at the State Theatre, in Easton. We're in the heart of downtown Easton, and there's a lot of traffic at this point, and there's a lot [of] people going past in the daily shuffle. • Lafayette College campus—walking up the steps from the Bushkill Creek to Lafayette. I see the Civil War monument dedicated to Lafayette College students from the classes of 1839 to 1862. • Oh, hi. I'm on South Side, and I see a bunch of houses being built, and [unintelligible] and taking pictures, and a beautiful home that's . . . • I'm in the Crayola store on the intersection of Northampton and Third Street, in Easton, and what I'm looking at is wall after wall of Crayola-related merchandise. • Easton, Pennsylvania—Lafayette College campus: Enterprise Rent-a-Car, an orange cable crossing the street, hiding behind some cardboard boxes, wood, a tree, green leaves . . . • The 25th Street K-Mart parking lot with a Dixie Cup. It's rush hour, but we're not rushing. The end. •

I'm here on the bridge across Bushkill Creek, on North Delaware Drive, and I was looking at the rust on the rail, because this was all underwater in April. There is lots of traffic. And T & H Automotive will get my car serviced. • I'm now at Braden Airport Academy, and there is not much going on here. Just closed the garage doors for the airplane, and I haven't really seen any take off, except for one. • I'm looking head-on at a Dodge Caravan. They have a yellow air freshener in the shape of a pine tree on their rearview mirror. • Hi. I'm calling from Braden Airport Academy, and we can only get so far with where we can walk here because of security. Now, you need an escort to get around, and to go visit the planes, and look inside the planes. When the kids were little, about five years ago—probably before September 11th—we could roam this property and look in the planes. • Whitman Reporter number 74: I'm sitting atop Mount Ida, which is on the

south bank of the Lehigh River where it meets the Delaware River. I have the vista of the entire downtown of Easton, Pennsylvania, in front of me. Behind me is the main railroad artery, which connects Easton with New Jersey and New York, and I think you can see from the video that I sent that that is now pretty desolate. • I'm on a two-lane paved road underneath a four-lane Route 22 overpass. Down here, where it should just be the crickets and me, there is traffic. There's an old apple tree, a pear tree, and it looks like a dead lilac—one that's been . . . • I'm standing at the corner of Third and Larry Holmes Drive, in downtown Easton, a pretty busy intersection this time of the day with people coming home from work. I see the sign that's pointing to Philadelphia that's telling me it's 58 miles away. Lots of people going in and out of McDonald's, and Wawa, and the Easton Inn. • Easton Parking Garage, looking at all the police vehicles. The canine vehicle: It's black,

really long. I guess they have to have room for the dog. It says, "Stay back," on the window. There is a canine trooper car, another canine vehicle; I guess they're all canine vehicles. No, here's an unmarked car. • We're at the 25th Street exit, where, umm, the Dixie Cup is; it was a landmark of Easton for a very long time. • Centennial Park, 12th and Ferry Street: getting ready for the closing celebration of the park. There is going to be a race-car driver later, and all kinds of celebrations and eating, and that's all we have for . . . • We're on the corner of Northampton Street and Larry Holmes Drive, on the Pennsylvania side of the "Free Bridge," facing east. We're trying to get across the bridge, but the sidewalk's closed. The people try to get to the other side. Will she be able to do it or won't she? The sidewalk is closed; we cannot cross the bridge at this time. • I'm standing on [unintelligible], with blue shutters and dirty windows. There is a green shark in here and there

is a yellow ladder and a box that says . . . • We're at Eddyside Park here, in northern Easton. When we first arrived at the pool, nobody was there, except seven lifeguards standing around smoking cigarettes and waiting for people to show up. Now, a couple of families with little kids have shown up, and they're heading for the kiddies pool. There is also a large hawk sitting in a tree. • I'm watching a man in a burgundy T-shirt, with jeans and boots on, spraying a tank with white spray paint. He's walking over to a burgundy car. Actually, he left the tank, telling us we should move. • I am now sitting on a thingy that is across from a brick building that is called [unintelligible] Furniture, and we're at the Braden Airport, and the skies are very dark. The top of it is kind of shiny bright: The sun is over there. And there's not many cars in the parking lot, and the airplanes aren't really taking off right now. • I'm watching a number of geese cross a little tiny stream. I guess there are

seven of them, and I'm looking at a cornfield and the area of the cornfield that is plowed. There is a young man who is actually hitting golf balls. And I see lots of sleeping cats—golden colored, black colored. • I'm in Easton Cemetery, right across from the largest of the mortuary temples: It's rough-cut stone, 30 feet tall. When you look inside, there is a Tiffany stained-glass window you can see at the far end, beyond the tombs, that's been smashed. It's at the very end of the cemetery, and looks both Egyptian and Roman at the same time. • The location is Easton Area Middle School, Cottingham Stadium—12th and Northampton Street. And, umm, there was a mural done there this summer by some local teenagers. • I'm on the campus of Lafayette College, walking, watching the food-service workers put up the tables for tonight's dinner. • I am looking, and I'm in an alley in back of South Main Street, Philipsburg, and I am looking at sparrows in a pear tree—not a

partridge, sparrow—from which Julie has just removed one piece of fruit. • There are tons of planes here, and they are almost all white, but some of them have different colored stripes on them. Most of them do: green, blue, royal blue, red, aqua, teal. • I'm watching the Lehigh River weave its way through a concrete canal, making its way into the Delaware River, with a little girl next to me throwing chocolate chip cookies into the canal. And her mom is smoking a cigarette, and it looks like she just got off of work. She works at the hospital. And I'm going to move away because the sound, the sound of the . . . • I'm in my own apartment, on South Fourth Street, because I got hungry and I needed to use the bathroom. • I'm in Nazareth, Pennsylvania. It's a light wind here blowing the American flag and the flag of Pennsylvania. There is a cannon here: black, cast-iron cannon donated by the War Department, and it's commemorating the veterans of Nazareth, Pennsylvania, who

fought in the Civil War. • I'm near the Circle, and I currently just ran into my friend Sean, who I haven't seen in about two years. I'm looking at the bench, and there is a guy with an amazingly awful checkered, plaid, orange-and-blue shirt, and the most massive cowboy hat I've ever seen. • "Private Property," "Keep Out," "41-12 Hexville Road," "Warning," "No Trespassing," "This Is Private Property," "Criminal Trespass." These are all the signs that are on the gate to this long driveway that leads to a large house with horse stables. • [Unintelligible] . . . is silver and black phone, and I am watching my friend sitting on the green grass and she is . . . • Above me, two red-tailed hawks are [unintelligible] to each other. On the ground, I'm at Lower Wade Road and Clover Hollow. • The location is Northampton Street around Fifth Street, where the State Theater is, and a tattoo place, and Rock Church, and a lot of people walking around. • I'm on the campus of Lafayette

College, watching students of the class of 2009 get ready to take their class picture. • Whitman Local Reporter Number 74 atop the Third Street railroad trestle in Easton. I'm overlooking traffic in downtown Easton, which is about average for Friday afternoon rush hour. People don't look too unhappy. • Hello. Am I reporting? I've been recording. O.K. I'm in [unintelligible], some guy in a convertible just went by. I'm sitting here with Mitch Jaeger and his wife [unintelligible]. She's got the wind in her hair and the south at her back; and there's a flag flying and a light breeze. • We are at a bar across the street from Center Square, Easton. There are about 30 people in here, and they're all drinking Labatt's Blue from the glass. • I'm here at the [unintelligible] Easton again on Third and Pine streets, where I can see a historical sign telling me that, where I'm standing, in 1846 there was a gunsmith shop, and that Samuel Philip was the owner, and he invented the bamboo fishing rod.

• There's a beautiful cast-iron water fountain here in Nazareth. It says, "Made by the Murdoch Manufacturing and Supply Company, Cincinnati, Ohio." But here it is in Nazareth, Pennsylvania; and if you step on the pedestal, water comes out. Listen—can you hear that? Mmm . . . • (Is it JonBenet Ramsey?) [Unintelligible] Auto parts store. I don't know if this went through. Sorry, give me one second. • Oh, I'm recording. I'm standing on concrete, and there are metal poles sticking out of it, and the sun is just peeping through the dark blue clouds. • I'm standing in the Maple Leaf Dairy Farm parking lot, just north of Lehigh Valley Industrial Park. "The Moo's Palace" is written on the side of a corrugated structure that has collapsed. The roof is falling down, pitching towards the right. The sun is setting in the background. There is a propane tank in the foreground. The parking lot is empty. The cornfields . . . • I see the Social Security building, a lot of traffic. There's a goddamn

American flag, and there's an old guy sitting on a bench. He's not doing anything. • There are these poles on this building, and they have snakes wrapped around them. The snakes are pretty long, and they're dead. • I'm walking along this street, not too far from the Circle, in Easton, and I'm under a tree that's dropping a lot of acorns. I don't know why the acorns are dropping so soon; it's not even September. • There's a rope coming out of a hole in the ground. • Easton, Pennsylvania—Lafayette College, Quad Drive: parked cars on the side of the street, and Marilyn Monroe leaning against a tree. • We're on North Delaware Drive watching the traffic go by, and a group of runners from Lafayette just ran past us. We're looking at a stone wall, and flowers, and . . . • A man has just loaded his minivan with 20 or so sacks of groceries right outside the local Wal-Mart Supercenter, right here in America. • I'm standing at the corner of Church Street and North Third

Street, in Easton, Pennsylvania. And it's about 20 after five, and it's a beautiful blue-sky day with puffy clouds, and the traffic's going by. • I am sitting on a curb, which is brown, and I'm across from a car dealership named Scott's. There is a stop sign across the street from me. There is a red car going . . . • My dog is wandering down a path along the Lehigh River, and going up to a man that appears to be homeless, who takes a small hit out of something that appears to be a roach. My dog is frustrated because she can't swim in the . . . • (Just go away, go away. Go away; I don't want to be distracted.) I see a stop sign. I see a blue garage with three doors. I see a John Deere tractor, and a Ford blue pickup truck. • I'm looking at: "Oprah quits," "Marilyn was murdered," [unintelligible], "Angelina talks," and a young woman wearing a silver belt, and two-tone hair, is reading People magazine in a bodega. • I'm walking towards my car on Third Street, in Easton,

137

making sure my meter doesn't run out 'cause there is a cop car behind me, and I don't want to get a ticket. And I can see I still have 17 minutes to go. • Looking through a garbage can on Third Street. She had a paper. There is a McDonald's cup in there. • I'm at the corner of Second and Ferry—one of the quiet corners around in the city today. The most activity we're seeing: two horse-drawn carriages coming just down the street. • The location is Bank Street and Northampton, at the Weller Center, where all the buses go in, and that's about it. Bye. • I'm sitting here between the Carmelcorn Shop and Julie's Emporium, and I'm watching this man put together the absolutely beautiful pieces of antique furniture, including this purple bench, and this wonderful leaf table with heart oak. The front of the store is gorgeous; there are all these stained-glass pieces and mirrors. • I'm on South Main Street, in Philipsburg. There is a guy in the trunk of his truck—no, his

car—and he's got a lot of wrapped packages: a fresh batch of cigars, Kool cigarettes, other things like that. He's taking out about four boxes and walking down the street into the . . . • I'm at the bus station, the bus terminal, and there are three dudes sitting on a bench. One of them has a huge Afro, and here comes the . . . • Whitman Local Reporter Number 74 standing on the Third Street railroad overpass. A westbound freight train just passed from New Jersey into Pennsylvania. Traffic is still relatively mild in town. • We're leaving Braden. We're leaving Braden Airport now, and we're about to make a left at the light on the corner. The Industrial Park is to our right, and we didn't see a lot action here today; not a lot of flying. • We're driving back to Wal-Mart now, Wal-Mart Plaza at Northampton Crossings, with a . . . • (Sit. Sit.) My dog is frustrated she can't go into the river, because the Delaware River was contaminated this . . . . Hi. I was on my way to Klein farms, but I've stopped

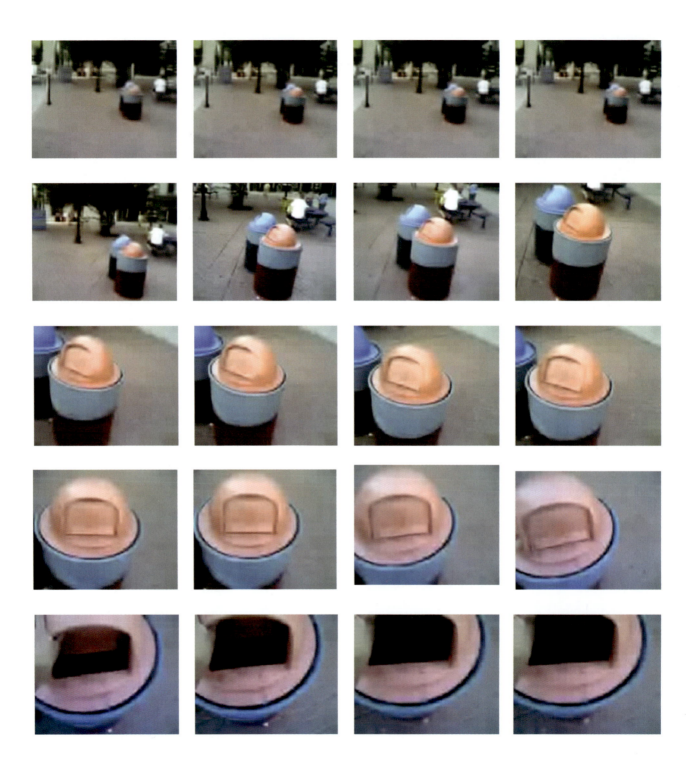

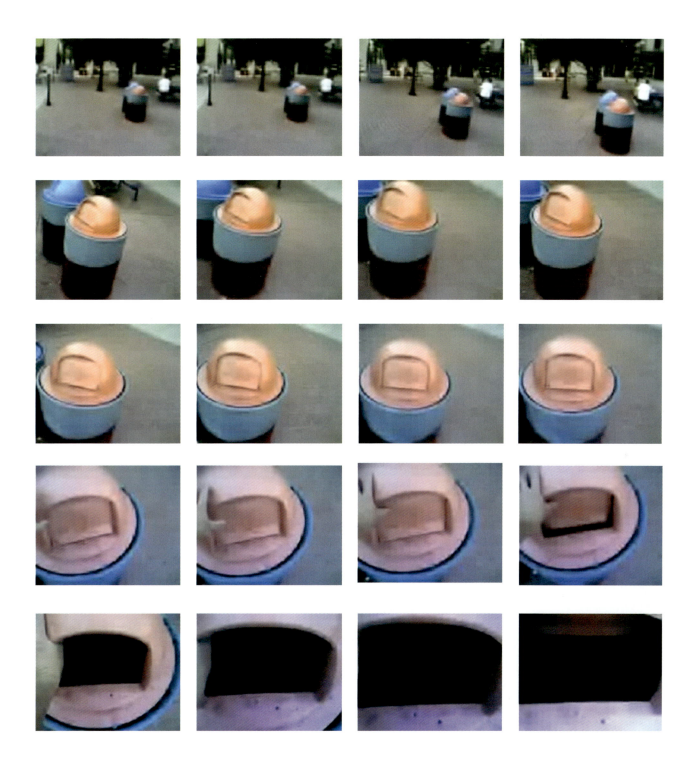

Instructions for the Local Reporters

Robert Whitman

There are two parts to the performance: first, making the news reports; then, showing the final sound and video work made up of these reports. The section for making the reports lasts approximately 30 minutes. The final work will be recorded and shown immediately, and for several more days, at the shopping center site.

The performance will also be streamed live to the project's web site: http://www.whitmanlocalreport.net.

Local Reporters should arrive at the shopping center performance area at 3:00 p.m., two hours before the performance.

Each reporter or reporting team will sign out his/her video cell phone and will take detailed maps and directions to the three specific locations from which they will report.

The reporters will be given training on the use of the phones for voice and video. The phones have been specially programmed for the project, to simplify use and minimize the number of buttons to be pushed. The reporters will be able to practice all the commands and functions and will have enough time to get comfortable with their phones. In addition, the reporters will be

Robert Whitman talking to Local Reporters, at Hawley Lane Plaza, Trumbull, Connecticut.

provided with written instructions on the use of the phone to take with them to their reporting locations.

Bob Whitman will talk informally about the process of reporting.

Reporters will leave for their first location around 4:30. The reporting will run from 5:00 to 5:30.

After each reporter or reporting team has made their three reports, they are to return to the shopping center performance site for a reception, where they can view the full work they have participated in creating. The reception will start about 6 p.m.

The sound and video will have been recorded, and the resulting video and sound work will be looped and will play continuously at the center for the next two days.

Everyone is invited to attend *Local Report* during its repeated presentations.

Score or Script for the Performance

Each Local Reporter goes to three sites and makes voice and video reports from them during the 30-minute performance time.

The timing of the calls and time between calls is broken roughly into five-minute segments. The schedule for one reporter, for instance, is to make calls and transmissions in the first five minutes. They take and send video of something they see first, then they call and make a verbal description of something else they see. After they have completed this, they move on to their next site during the next five-minute period, and make video transmissions and calls in the third five-minute period. They repeat this and make their final transmissions/ calls in the fifth five-minute period.

The timing of the calls will be staggered so that, say, one group of 15 reporters will make their transmissions/calls during the first five minutes, move on, and make calls/ transmissions in the third five-minute period, and make their final calls/transmissions in the fifth five-minute period. The reporters in the second group of 15 will make transmissions/calls in the second five-minute period, move to the next location and make calls/transmissions in the fourth five-minute period, and make their final calls/transmissions in the sixth five-minute period.

The variations in signal strength and transmission time for video uploads at Trumbell led to changing the timing for sending reports. These were the new instructions at the next four venues:

Local Reporters will leave the shopping center site to go to their first reporting location at 4:00; they are to begin taking and sending video reports at 4:30. At 5:00, they will begin to make their voice calls. After the voice call, they will go to the new location and, first, make and send a video report and then a voice report, and then move to their next location. They may alternate video and voice reports until the piece ends at 5:30.

Video Reports

You may take and transmit your video clips at any time without waiting for a "line."

Press WLR key to activate the video-taking function on the phone. Take a 20-second video of whatever you want. Press the key to send the video.

Please stay where you have shot your video until the screen tells you that the video has been sent successfully.

Voice Calls

Press the key for voice calls.

You can call any of the five numbers programmed into the telephone's memory.

When you make your voice call, you may be put on hold and hear music.

There will be a voice telling you, "You're next."

You begin to talk when the music stops.

Begin your call by saying where you are—for example, "I'm on the corner of X and Y streets."

Describe something you are seeing.

The call will end when you hear a voice saying, "Thank you. You're off."

Then, you can move to the next location to make your next set of reports.

Your video clips and sound calls may describe different things.

At 5:30, return to the shopping center.

Robert Whitman talking to Pauline Oliveros, who was a Local Reporter at Kingston, Center, Kingston, New York.

Local Reporter Robert Mattison and Walter Patrick Smith at Northampton Crossings, Easton, Pennsylvania.

143

How to Create Your Own Local Report

Shawn Van Every
Hans-Christoph Steiner

Introduction

It is very possible for anyone to run their own Local Report, and it does not take a large-scale production to make it work. Given time, patience, and resourcefulness, a moderately tech-savvy producer might be able to draw on the resources given here and set up the infrastructure for running Local Reports themselves. At a minimum, running a Local Report requires an Internet connection, Voice Over Internet Protocol (VoIP) service, a decently capable computer, and collaborators with video-capable mobile phones (to serve as reporters). Running the Local Reports, as documented in this book, required a relatively complicated computer setup, since there were many facets to the production. If you want to run your own Local Report, you need not include every aspect of the production to succeed. For instance, your Local Report could be played live on a screen and speakers, then recorded and posted to the web site later. In that case, you will not need to set up the streaming infrastructure. Or, you can have the event only displayed live via streaming, in which case you will not need local playback and recording capabilities.

Overview

The Whitman *Local Report* was run at five locations over the course of two months. At each location, we had an Internet connection, several laptops, a projector, and, of course, 30 "reporters" with mobile phones. The reporters were sent out into the local area with the task of recording video in 20-second snippets, which were sent back to the *Local Report* headquarters, and calling in live with a verbal report as well. At headquarters, we sequenced the video as it came in and played the audio live when the reporter called. We projected this and played the audio on large speakers for anyone to come and listen and watch. We

also streamed it out live into the Internet so anyone anywhere could watch.

Below is a breakdown of the individual pieces that comprised the project, presented with the intention of helping you produce your own version of the Local Report.

Mobile phones

For the Whitman Local Report (WLR), we used 30 Nokia 7610 phones. These phones support development using Java 2 Micro Edition (J2ME) and have additional capabilities for video and networking (MIDP 2.0 and

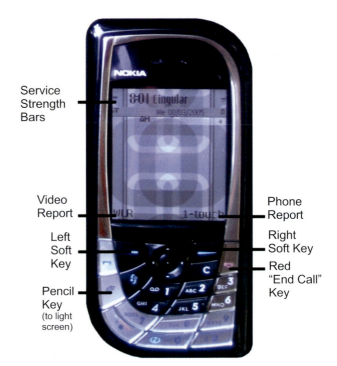

Diagram of the Nokia 7610 cell phone with identifications of programmed keys for the Local Reporters.

MMAPI). Nowadays, these capabilities are common on many handsets, and the application that was developed for WLR could be ported to a variety of other models. The basic requirements are a video camera plus support for J2ME (with MIDP 2.0 and MMAPI) with video capture.

The source code for the WLR phone application is available online at: http://www.whitmanlocalreport.net/phone/source/.

If you are interested in giving it a try before attempting to modify the source, the compiled application can be downloaded directly to your phone and tested by going to the following URL:

http:/www.whitmanlocalreport.net/phone/
   Whitman_Mobile_Video.jar.

There is one caveat to trying this application: The sending of the video to the *Local Report* server requires a data connection through your mobile phone service provider (T-Mobile, Cingular, Verizon . . .), and use of this data connection may cost extra, depending upon your service. Check these costs before attempting to use the application.

The source for the application can be compiled and installed as is, but unless you make some minor modifications, it will upload the video to the Whitman *Local Report* server instead of where you may wish it to go. The upload URL (HTTP POST) is hard coded into the source.

To change the upload location in the Java source you will need to modify the following line in the Whitman-MVideo.java file and input the correct URL in place of the www.whitmanlocalreport.net/phone/up.php URL:

HTTPPoster poster = new HTTPPoster
   ("http:/www.whitmanlocalreport.net/phone/
   up.php", "post.3gp", videoBytes);

If you try out the application without modifying the source, you should be able to find the video you upload by going to: http://www.whitmanlocalreport.net/videos/.

Last, this application has a couple of features/bugs/items on the to-do list. We welcome any modifications or contributions back to the source if you are interested in helping out.

1: The status indication for uploads does not accurately indicate the progress of the file upload. This is unfortunately a limitation in the Java platform on mobile devices.

2: The videos are not saved on the device. There is an attempt to utilize a "record store" to do this, but it's implementation is not complete and does not function correctly.

Again, the caveat: This application requires the use of a data connection on the phone from your mobile provider. If you do not have a plan to encompass data use, you will generally be paying a premium for the service. Second, video files are generally fairly large and can eat up your allowed data usage. In other words, BE CAREFUL with how you use this until you know what it will cost.

Uploading the Videos

The mobile phone application uses HTTP file upload capabilities to send the video to a web server on the Internet. The very basic (and very insecure) PHP script can be used to receive the video files sent from the phone:

```
<?
$uploadfilename=time()."_".rand(1000,9999)."_".
   basename($_FILES['bytes']['name']);
$uploaddir='/xxxx/xxxx/xxxx/videos/';
$uploadfile=$uploaddir.$uploadfilename;

if (move_uploaded_file($_FILES['bytes']
   ['tmp_name'], $uploadfile))
{
   echo "Video Sent!";
}
else
{
   echo "Error: Video Not Sent!";
}
?>
```

In order to use the above script, you must put the code in a text file with the extension ".php". You must also change the line that starts with "$uploaddir" to point to the location to which you want the video files saved on your server. You will need a web server that can run PHP scripts, and this script will have to be placed somewhere that is web-accessible (you probably want it in the location that you are pointing to with the mobile phone application).

Video Playback Application

On-site, at each location, we had a laptop hooked up to a projector that ran an application to play back the videos full screen as they came in from the server. This application is actually two applications. The first is called "rsync," which is an application for synchronizing files across a network. Rsync was running on the server as a "daemon," meaning that it was running as a service in the background. It was configured to share the videos "read-only" to any connection, so that we did not need to set passwords or have to worry about someone deleting or moving the files. Rsync, then, also ran on the video playback laptop to download the incoming videos from the server. It was set to run using "cron," a scheduling service that ran the rsync program every few minutes to download any new videos.

The second portion is a Java application that uses QuickTime for Java (which is bundled with QuickTime and iTunes) to play back the videos in sequence, with controls for playing the next and previous videos. Although this application only plays 3gp videos, it can be modified to play other QuickTime-compatible formats fairly easily.

The source for this application can be downloaded at:

http://www.whitmanlocalreport.net/video_player/.

On-site Computer Setup

At each location, we had an Internet connection set up and ready to go when we got there. We then set up a small network of computers to run the show. One laptop was dedicated to managing the incoming calls and playing the audio from those calls. The next laptop was used for downloading the videos from the server playing them back. Both the audio and video were sent to another laptop that combined them into a stream, which it then sent on to the streaming server. The audio/video signal was also sent to a local projector and PA system for live playback on-site.

Server Setup

In addition to the collection of laptops used on-site, there was one remote server. The server, running the Debian operating system, was located at a colocation facility in Los Angeles. It provided a number of essential functions: web site hosting with the Apache web server; phone-call termination and call-queue system using Asterisk phone-system software; custom software for receiving videos sent from the phones and relaying them to the laptops on location; and the public streaming server, running Darwin Streaming Server, that delivered the audio/video stream for the world to experience.

Phone System (Asterisk PBX)

The central part of the voice reports is the phone system built using Asterisk. It is, essentially, a call-queue system, such as any business might have when you call its customer-service number. Participants dial one of the phone numbers preprogrammed into the mobile phones. Asterisk answers, plays back some basic instructions, then moves on to the hold music, announcing the caller's position in the queue whenever a call is answered. The first caller in the queue is transferred to the extension of the software phone running on the laptop on location. This software phone acts just like a phone in an office phone system; it has an extension number that works only within that context. When the participant is next in the queue, Asterisk plays a prompt to get them prepared to deliver their report: "Get ready, you're next! When the hold music stops, you are live on the air!" After the current call hangs up, the next participant is prompted, and Asterisk sends the next call to the software phone on location. The software phone was configured to answer automatically as soon as possible, putting the audio immediately on the air. When the current report hits a good breaking point, hit the Esc key on the laptop running the software phone. Then, the call is dropped and the next one picked up, starting this process again.

To set up your own queue system, you will need an Asterisk server (http://www.asterisk.org) and VoIP. You can find lots of assistance for setting up Asterisk help online, so this is just an outline of what needs doing. It is probably worthwhile to get a hosting provider to install and host Asterisk for you. Once you have Asterisk running and receiving calls from normal phones (mobile or otherwise), you can set up the call-queue mechanism. All of the queues are set up in a file called "queues. conf". In this file, you can set up the various automated announcements, the frequency of the announcements, and which extension receives the calls from the queue. All in all, the *Local Report* setup only required about 10 custom settings to get the call queue working.

Next, you will need a piece of software called a "soft-phone" to receive the calls as they come in. For this purpose, we used software called Freshtel Firefly, but there are many solid options available now, depending on which operating system your computer is running. Essentially, this software works just like a phone, receiving calls using VoIP instead of a normal phone line. We hooked this phone into Asterisk to serve as the phone that Asterisk directed the calls to once they were in the top of the queue. Again, setting this up is out of the scope of this document, but plenty of information for configuring soft-phones to work with Asterisk can be found online.

Last, Asterisk may be one of those portions of the system that we used for our setup that may prove to be overkill for what you want to do. Many VoIP TSP (Telephone Service Providers) allow you to configure service with them to work directly with a soft-phone. Some may even allow you to set up a queue. We used Voicepulse for our service. They come highly recommended, but the industry is changing quickly, and it would behoove you to shop around a bit.

Mobile Phone Configuration

Our mobile phones were configured to make them as easy to use as possible. We wanted people with no mobile phone experience to be able to participate fully in *Local Report*. Much of this configuration was centered around making the user need as few button clicks as possible to make their reports. In particular, we configured the two "soft keys" to open the two relevant applications for *Local Report:* the video application and the contacts (for calling in).

Kieran Sobel setting up the outside speakers for the performance at Liberty Square Center, Burlington, New Jersey.

Patrick Heilman installs the sound equipment as Shawn Van Every and Hans-Christoph Steiner set up for the performance at Hawley Lane Plaza, Trumbull, Connecticut.

When a participant clicked on the soft key to launch the video application, the user needed only to keep clicking on that same soft key to choose all of the requisite options to record a video and upload it to the server. The first click launched the application, the second click started recording video, the third click stopped the recording, then the remaining clicks uploaded the video. When a participant clicked on the audio soft key, they were presented with four different phone numbers to dial, all of which dial the same phone system. The participant just clicked once more to dial a number, then followed the voice prompts of the phone system to do their audio report.

Making a Video Report

First, the participant launched the video application as described above. This opens a preview window to set up the shot. Following that, a press to the video soft key (the left button) starts the recording. Once recording starts, the border around the video preview turns red, and a countdown from 20 begins, showing the remaining recording time. The participant may compose the video

Robert Whitman, Shawn Van Every, Kieran Sobel, and Walter Patrick Smith at the computer control station during the performance at Kohl's Plaza, Holmdel, New Jersey.

however they wish within the 20-second limit. When the countdown hits zero, the participant is prompted to upload the video, which is done by a couple of clicks on the video soft key (allowing the application to access the network). Once the video upload is complete, the participant is notified on the phone, and the video is automatically played on-site and on the stream.

Making an Audio Report

The Asterisk phone system answered the call, played a prerecorded welcome, then placed the call in a hold queue, complete with hold music. As the call advanced in the queue, the phone system announced the call's position in the queue. When the call was up next, the participant heard a little pep talk, getting them ready to deliver their report. When the call was finally answered, the hold music stopped and the participant was immediately speaking live on location and the stream.

There are indeed many parts that make up the whole system that runs a Local Report, but each of the pieces is relatively simple in the grand scheme of things. Fortunately, there is a lot of good, free software—like Asterisk, rsync, and J2ME—that this whole system is built upon. Using the software provided on the whitmanlocalreport.net web site, your own *Local Report* is that much closer to happening.

Local Report Participants

Performance:
Robert Whitman
Shawn Van Every
Hans-Christoph Steiner

Project Management:
Walter Patrick Smtih
Julie Martin

Production:
Patrick Heilman
Anne-Olivia Le Cornec
Paul Miller
Martin Palacios
Sven Schroeter
Kieran Sobel
Hedi Sorger

**Post Production of
Audio and Video:**
Walter Patrick Smith

Hawley Lane Plaza
Trumbull, Connecticut
29–31 July 2005

Local Coordinator:
 Cynthia Beth Rubin

Kelly & Rowland Becerra
Luke Bellwood
Steve Bellwood
Jim Berger
Anna Bresnick
Kat Burns
Vicki Cain
Bishen Canth
Sebastian Castano
Jeanne Criscola
Dan Fisher
Joan Fitzsimmons
Yolanda & David Goff
Susan Greenberg
Ted Gutswa
Doug Jacoby
Gregory Jacoby
Jeff Jacoby
Jennifer Klein
Janet Lehmann
Eudald Lerga
Carol & Joel Leskowitz
Harry J. Martin Jr.
Yolanda Petrocelli
Jeanne Quickmire
Philip E. Rubin
Suzan Shutan
Christina Spiesel
Sharon Steuer
Colleen Tully
Heather Van Deusen
Ben Westbrock

Kohl's Plaza
Holmdel, New Jersey
5–7 August 2005

Local Coordinators:
 Eileen Kennedy
 Terri Thomas

Lisa Agresta
Andrea Barauskas
Derickson Bennett
Robyn Ellenbogen
Zephan Ellenbogen
Amy Felton-Toth
Paul Guba
Harry Hukkinen
Maria Milazzo
Marygrace Murphy
Donna Payton
Dennis Reynolds
Max Roberts
Tara Durkin Rochford
Anthony L. Rodriguez
Devon Rust
Susan Smith
Kaitlin, Griffin & Aislinn Smith
Nina Sobell
Tony TeDesco
Alizabeth Toth
Steven Trezza
Rita Williams
Zack Williams

Liberty Square Center
Burlington, New Jersey
12–14 August 2005

Local Coordinators:
Julia Robinson
Sid Sachs

Peter Cho
Gabriela DiVentura
Carter Farmer
Jen Goettner
Ryan Gother
Rudy Guenther
Aaron Igler
Donghee Ma
Alexis Macrina
Julie Martin
Andi Monick
Christine Murphy
Dede Muse
Robert O'Connor
Martha Perry
Patti Repetto
Scott Rigby
Margaret Robinson
Joseph Rossi
Rena Segal
Angela Smith
Wade West
Isaac Witkin

Kingston Center
Kingston, New York
19–21 August 2005

Local Coordinator:
Lisa Barnard

Kathleen Anderson
Jim Barden
Sylvia Bullett
John Dubberstein
Betty Greenwald
Steve Hopkins
Kyle Kelley
Michelle Kuo
Jenny McShan
Seth Nehil
Pauline Oliveros
Tasha Petersen
Jacob Proctor
Hedi Sorger
Marjorie Strider
Graham Taylor
Eric Vigil
Dennis Warner
Cynthia Whitman Stoll
Pilar Whitman
Bethany Wright

Northampton Crossings
Easton, Pennsylvania
26–28 August 2005

Local Coordinators:
Michiko Okaya
Robert S. Mattison

Kristen Barnes
Bill Deegan
Letty Lou Eisenhauer
William Fetterman
Patricia Goodrich
Daniela Hellmich
Polly & Dick Kendrick
Krista Laubach
Claire & Faye Lukas
Helen Martin
Julie Martin
Ana Mattison
Liza Mattison
Robert S. Matison
Spencer Mattison
Owen McLeod
Mia McPeek
Paul Miller
Yvonne Osmun
Martha Posner
Rebecca Price
Joe Shieber
Demetra Stamus
Shawn Van Every
Wendy Van Every
Rhonda & Bruce Wall
James Weiss
Mary & Bill Zehngut

Biography of Robert Whitman

Robert Whitman was born in New York City in 1935. He studied literature at Rutgers University, from 1953 to 1957, and art history at Columbia University, in 1958. He began in the late fifties to present theater performances in spaces in downtown Manhattan, and was the first artist to incorporate film into such pieces. These works, rich in visual and sound images, incorporate actors, film, slides, sound, and evocative props in environments of his own making. His seminal works include *American Moon, Flower, Night Time Sky,* and *Prune Flat.*

Whitman has presented more than 40 theater pieces in the United States and abroad, touring to various European venues, including the Moderna Museet, Stockholm, and the Centre Georges Pompidou, Paris. He has had one-person exhibitions of his sculpture and installation pieces at numerous museums, including the Jewish Museum, New York; Hudson River Museum; Museum of Contemporary Art, Chicago; Museum of Modern Art, New York; Thielska Galleriet, Stockholm; and one-person gallery exhibitions at PaceWildenstein, in New York.

In 1966, Whitman was one of 10 New York artists to participate in *9 Evenings: Theatre & Engineering*, a series of performances at the 69th Regiment Armory, New York, presenting "Two Holes of Water-3."

Whitman was one of the cofounders of Experiments in Art and Technology (E.A.T.)—along with engineers Billy Klüver and Fred Waldhauer and artist Robert Rauschenberg—to provide the contemporary artist with access to new technology as it developed in research institutions and laboratories.

He has collaborated with engineers on installations and works that incorporate new technology: laser sculptures, including *Solid Red Line* and *Wavy Red Line*; *Pond,* a sound-activated Mylar mirror installation shown at the Jewish Museum, New York, in 1969. His long collaboration with optics scientist John Forkner began with a mirror, light, and sound installation for the exhibition *Art and Technology,* 1971, at the Los Angeles County Museum of Art.

Whitman was one of the core artists for the Pepsi Pavilion at Expo '70, Osaka, Japan, a project administered by E.A.T., and was instrumental in designing one of the main features of the interior of the central performance space, a 90-foot diameter 120-degree spherical mirror made of aluminized reflective Mylar, which produced real images of the visitors hanging upside down in space.

Whitman, working with E.A.T in the early seventies, developed and participated in a number of innovative communications projects: *Anand Project, Children and Communication, Telex: Q&A,* and *Artists and Television. Ghost,* a recent theater performance, was staged at PaceWildenstein Gallery, October 2002; and *Antenna* was presented at the Leeds New Media Festival, sponsored by Lumen, in October 2004.

In 2003, Dia Center for the Arts, New York, presented *Playback,* a large-scale retrospective exhibition of Whitman's works. The exhibition traveled to Museu d'Art Contemporani de Barcelona (MACBA), in 2004, and to the Museum of Contemporary Art at the Serralves Foundation, Porto, Portugal, in 2005.

Whitman collaborated with Walter Patrick Smith and Roger Duffy, of SOM Education Lab, on the design for the Camden Creative and Performing Arts School, in Camden, New Jersey, which in 2004, won an award from the AIA New Jersey. He is currently collaborating on the design for a performing arts center at the Brunswick School in Greenwich, Connecticut.

Whitman has received many awards, including a Guggenheim Fellowship; Creative Artists Public Service Grant; Citation of Fine Arts, Brandeis University; and Creative Arts Award Xerox Company Grant.

He is represented by Pace Wildenstein, NY.

Selected Bibliography

9 Evenings: Theatre & Engineering. New York: Experiments in Art and Technology, 1966.

Cooke, Lynne, Karen Kelley, and Bettina Funcke, eds. *Robert Whitman: Playback*. New York: Dia Art Foundation, 2003.

Kirby, Michael. *The Art of Time: Essays on the Avant-Garde*. New York: E. P. Dutton, 1969.

———. *Happenings: An Illustrated Anthology*. New York: E. P. Dutton, 1965.

Klüver, Billy, Barbara Rose, and Julie Martin, eds. *Pavilion*. New York: E. P. Dutton, 1972.

Kostelanetz, Richard. "Interview with Robert Whitman." *The Theatre of Mixed Means*. New York: Dial Press, 1968.

Marter, Joan. *Off Limits: Rutgers University and the Avant-Garde, 1957–1963*. New Brunswick, N.J.: Rutgers University Press, in association with Newark Museum, 1999.

Rose, Barbara. "Considering Robert Whitman." *Projected Images: Peter Campus, Rockne Krebs, Paul Sharits, Michael Snow, Ted Victoria, Robert Whitman*. Minneapolis: Walker Art Center, 1974.

Van der Marck, Jan. *Robert Whitman: 4 Cinema Pieces*. Chicago: Museum of Contemporary Art, 1968.

Whitman, Robert. *Palisades: Robert Whitman*. Yonkers, N.Y.: Hudson River Museum, 1979.

Contributors

Bettina Funcke's first book, *Pop or Populus: Art between High and Low,* is scheduled to be published by Walther König. The New York-based writer is a member of Continuous Project, a collective that publishes books as well as produces installations and performances. Currently, she is working on a project that the collective considers its tenth issue; it will be both a performance and the catalogue for an upcoming exhibition at Modern Art Oxford. Funcke is also an editor at Dia Art Foundation, where she has coedited numerous books, including *Robert Whitman: Playback.*

Anne-Olivia Le Cornec received a doctorate in media/digital systems from CNAM, in Paris. She is now working on documentaries, including an interview film, with Billy Klüver, on E.A.T. She is also developing Babel Lexis, an interactive web site for children.

Julie Martin first worked with Robert Whitman as production assistant, in the summer of 1966, when he produced theater pieces at New York's Circle in the Square. She has continued to work with him on his own performances as well as E.A.T. projects. As coeditor with Billy Klüver and Barbara Rose of *Pavilion,* she documented E.A.T.'s building of the Pepsi Pavilion at Expo '70, in Osaka, Japan. With Klüver, she wrote *Kiki's Paris* and edited and annotated *Kiki's Memoirs.* Currently, she is director of E.A.T. and executive producer of a series of 10 video films on the artists' performances at the 1966 series *9 Evenings: Theatre & Engineering.*

Walter Patrick Smith is a senior designer and associate partner at Skidmore, Owings & Merrill, LLP, the New York architecture firm. Smith has designed traditional housing for the homeless and developed master plans for cities in China and Europe, under the direction of David Childs. He was senior designer of the new Terminal E at Boston's Logan Airport and the founder, along with partner Roger Duffy and Chris McCready, of the SOM Education Lab, whose mission is to elevate school, museum, library, and theater design by collaborating with individuals in many areas of the arts. In addition to working with Robert Whitman on a performing-arts high school in Camden, New Jersey, recent projects include collaborations with James Turrell on a library at Greenwich Academy, in Connecticut, Lawrence Weiner on a high school in Elizabeth, New Jersey, and Robert Irwin on a cognitive sciences building at Brown University, in Providence, Rhode Island.

Hans-Christoph Steiner is currently technical director of the Integrated Digital Media Institute, Polytechnic University, in Brooklyn, New York. Fusing three disciplines, Steiner composes music with computers, builds networks with free software, and works on interactive art projects. With an emphasis on collaboration, he has produced responsive sound environments, free networks that help build community, musical robots that listen, and interactive sculpture using homemade jets.

Shawn Van Every is adjunct professor and media researcher at New York University's Interactive Telecommunications Program, where he specializes in the development of participatory media. He works with a variety of platforms, from mobile devices and set-top boxes to large-scale servers. His projects generally involve emerging technologies and development of tools that help make it possible to produce and distribute both low-cost and interactive media.

Wendy Van Every, Local Reporter, in Easton, Pennsylvania, using her video cell phone to make video and audio reports.

Production Information
Transcription: Terry Martin, Meredith Snow, Hedi Sorger
Design Layout: Michiko Okaya, Julie Martin, Hedi Sorger, Gayle F. Hendricks
Cover Design: Pilar Whitman, Ben Bloom
Copy Editor: Gerald Zeigerman
Printed by Harmony Press, Easton, Pennsylvania

Photo credits
Vicki Cain, pages 13, 142
Courtesy Lumens, page 21
Julie Martin, pages 9BL, 10, 11, 12, 18BL, 143TR, 147 BL
Michiko Okaya, page 19BL&BR
Martin Palacios, pages 6, 9TR, 14, 15, 16, 17,19T, 27, 143BR, 147BR, 148
Peter Poole, page 24
Harry Shunk, page 20
Shunk-Kender, page 23
Walter Smith, page 18TR
Hedi Sorger, page 144
Shawn Van Every, page 155
Eric Vigil, page 29
Maps, copyright © American Map Company. Used with permission of
the American Map Company.

E.A.T.

Experiments in Art and Technology (E.A.T.) is a service organization that
promotes collaborations between artists and engineers and scientists.

69 Appletree Row
Berkeley Heights, New Jersey 07922
T (908) 322-5683

LAFAYETTE

Founded in 1826, Lafayette College is a private, coeducational institution
offering bachelor's degrees in the liberal arts, sciences, and engineering to
2,300 students from across the U.S. and around the world. www.lafayette.edu

Williams Center Art Gallery
Morris R. Williams Center for the Arts
Lafayette College
Easton, Pennsylvania 18042-1768
T (610) 330-5361
F (610) 330 5642
www.lafayette.edu/williamsgallery

The catalogue, *Robert Whitman Local Report*, can be ordered on line at www.whitmanlocalreport.net.